THE GREAT IRISH WEATHER BOOK

BY JOANNA DONNELLY

ILLUSTRATED BY FUCHSIA MACAREE

GILL BOOKS

THANKS!

JOANNA

First of all, I'd like to thank the people at Gill Books who brought this book to your table: Deirdre Nolan, who first introduced me to the idea of writing a book; Sheila Armstrong, our thorough editor who made great contributions to making such complicated material accessible to readers of all ages; and Teresa Daly and all the team at Gill without whom there would be no book.

Thanks also to the incredible talent of Fuchsia whose work complements the science of the book so well.

I can't give enough weight to the thanks I have to give my family. My mother, of course, who is smarter than she knows and gave me so much wisdom. Harm not only helped me along the way as I shared each idea for the book with him, cross-checking every fact, he helps me every day by being my best friend, husband and father to my children. Speaking of my children, Nicci, Tobias and Casper were the first to read and enjoy this book. I thank them for all their work and support over the past year. But mostly I thank them for their love.

FUCHSIA

For my parents and Ciarán.

CONTENTS

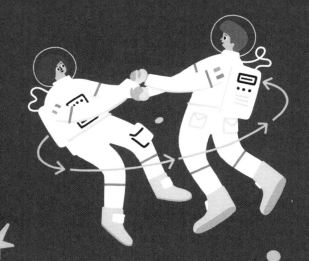

HAVE YOU MET SCIENCE?

PEOPLE OFTEN REMARK to me that I must have been good at geography in school. I wasn't. At first, I was confused as to why people said this – I thought maybe they were being sarcastic. I didn't even know where Timbuktu was (or that it was a real place – I now know, by the way!). But that's not why they asked. Apparently, it is because meteorology is part of the geography syllabus in the Leaving Certificate. I dropped geography after the Junior Cert because I wanted to do all the science subjects.

Science is the best thing ever! If you think you don't like science, well, that's probably because you just haven't been introduced properly. Science fascinated me as a child, and it fascinates me still. I think it fascinates many children, even if they don't know it. That's why children ask questions. 'Why is the sky blue?' 'Why does a ball bounce?' 'Why do I have to go to sleep?'

When a question requires a little more information than 'It just is!' or 'Because mummy said so!', scientists will go in search of the answers. The answer to why the sky is blue was researched by a man called John Tyndall. He discovered that the rays of light from the sun are scattered by the atmosphere.

Light is made up of lots of different colours, but when it is broken up on its way to Earth, we end up seeing mostly blue here on the ground. Balls bounce because of the elastic properties of rubber. And you need to go to sleep so your parents can get some peace!

But wait – is that last one true? It's an easy answer, but science isn't about quick or easy answers. It starts with looking at the world and coming up with an idea. We call this a theory. This theory may sound correct, but if we want to be sure, we have to test it with real-life examples. If our theory passes those tests, we can be certain that it is right. Sometimes it can take weeks, years or even centuries to find the answer. So, after a long time, and lots of tests, we now know that you need to go to sleep so that your brain and body can rest and recover. That's how proper science works!

These days, everyone has huge amounts of information at the tips of their fingers, and some people like to think they know all the answers. Sometimes you hear people say, 'Well, I'm entitled to my opinion.' But you don't want any old opinion – everyone needs to have an informed opinion. To get an informed opinion, you need the facts.

To get the facts, you need science – after all, it's called scientific fact, not scientific opinion! Science makes the world work, because we rely on these facts every day, in our hospitals, buildings, cars and food. To find the facts out for yourself, you can either ask people with expertise and experience – or you can do the work yourself, and become a scientist!

So, what does meteorology have to do with science? You probably know that it has something to do with the weather forecasts you see on the TV. But what it really involves is asking questions about weather. Why do we have clouds in the first place? How does the weather change all the time? Why do we appear to get so much rain and others don't? To answer these questions, we use the scientific method. So, meteorology really shouldn't be part of geography – it's science.

The type of science that we use in meteorology is called physics, which is the study of things like light, heat, sound, electricity and what things are made of – right down to the tiniest particles. We also use maths, of course, but then that's because maths is the language that science speaks. (But don't worry if you're not a fan of maths – you won't find much

in this book.) We mix physics and maths together to try to understand the weather.

Meteorology, like all science, starts with the simplest fact: 1+1=2. And we build on that. (Whoops, there's some maths. OK, there might be a little bit of maths!) And then we build on that again. We keep building on all the small facts until we have all the big ones, all the way up to the equations we need to forecast the weather. Some of our equations even look like this!

$$\frac{\partial V}{\partial V} + v.\nabla V = -\frac{1}{\rho}\nabla p + v\nabla^2 V - g$$

(That's all the maths for now, I promise.)

In this book, I'm hoping to give you some small facts, and these will give you an insight into meteorology. I hope that you can start to love it as much as I do, and maybe you'll decide to go to a university and find out all the big meteorological facts for yourself.

Now, let's go learn about the weather!

JOANNA

EXOSPHERE

THERMOSPHERE

MESOSPHERE

STRATOSPHERE

OZONE LAYER

TROPOSPHERE

METEOROLOGY

Meteorology is the part of science that looks at the physical processes that are going on in the atmosphere in order to understand the weather. The atmosphere is what we call the layers of air that surround the Earth and protect it from the sun. There are several layers, but all of the weather occurs in the bottom two layers, mostly in the troposphere. In this section, we will look at how that weather is created and different types of weather that occur in Ireland and all over the world.

THE SUN

If we want to talk about meteorology, we have to start with the sun. The sun is a giant ball of exploding gas located in the centre of our solar system, in the galaxy called the Milky Way.

THE SOLAR SYSTEM

The Earth and all the planets rotate around the sun. There are eight planets in our solar system – Mercury, Venus, Earth, Mars, Jupiter, Saturn, Uranus and Neptune – and the sun affects each of them in a different way. For example, the planet Mercury is closest to the sun, so it is boiling hot during the day. Neptune is the furthest away, so it's the coldest.

Earth is the third planet from the sun, so we are somewhere in between – just the right temperature for plants, animals and humans to live and grow. The sun is the source of all the energy on Earth. On it, hydrogen gas is exploding in a process called nuclear fusion. This reaction creates a huge amount of heat – so much heat that it reaches all the way to Earth.

RADIATION

The heat from the sun reaches us through a process called **radiation**. You've probably heard of radiation. I'm sure there are radiators in your home, and if you stand beside them when they're turned on, you can feel the heat radiating off them. You'll know about the sun's radiation too. When you're outside in the summertime, you have to wear sunblock to stop the harmful radiation from burning your skin. But not all of the sun's radiation is harmful. We need lots of it to warm the Earth so that we can live here. Light is also part of the sun's radiation – without it, we wouldn't be able to see anything!

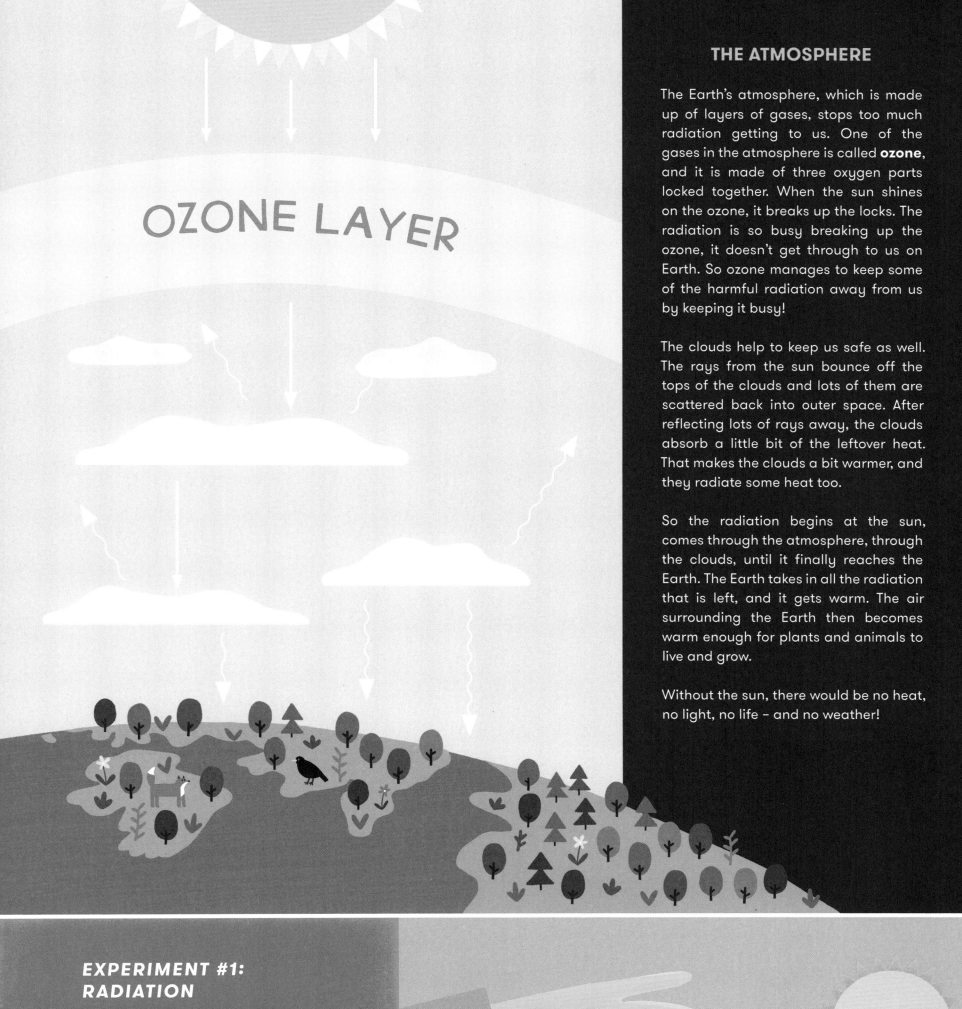

OZONE LAYER

THE ATMOSPHERE

The Earth's atmosphere, which is made up of layers of gases, stops too much radiation getting to us. One of the gases in the atmosphere is called **ozone**, and it is made of three oxygen parts locked together. When the sun shines on the ozone, it breaks up the locks. The radiation is so busy breaking up the ozone, it doesn't get through to us on Earth. So ozone manages to keep some of the harmful radiation away from us by keeping it busy!

The clouds help to keep us safe as well. The rays from the sun bounce off the tops of the clouds and lots of them are scattered back into outer space. After reflecting lots of rays away, the clouds absorb a little bit of the leftover heat. That makes the clouds a bit warmer, and they radiate some heat too.

So the radiation begins at the sun, comes through the atmosphere, through the clouds, until it finally reaches the Earth. The Earth takes in all the radiation that is left, and it gets warm. The air surrounding the Earth then becomes warm enough for plants and animals to live and grow.

Without the sun, there would be no heat, no light, no life – and no weather!

EXPERIMENT #1: RADIATION

On a sunny day, find a stone from the garden or the beach, preferably a dark one.

1. Leave it out in the sun. After it has been sitting in the sun for a while, bring it back inside.

2. Place it on the table and put your hand very near the surface. But don't touch!

3. You will feel the heat radiating from the stone, which has come directly from the sun.

THE EARTH

First things first: the Earth is round. Despite being able to dress themselves and turn their phones on, there are some people who are silly enough to think that the Earth is flat!

SHAPE AND SPIN

The Earth is a globe floating in space, but it's not perfectly circular. When the Earth was forming, billions of years ago, it was made of molten rock. Something called the **centrifugal force** meant that its middle bulged out, making it look like it was squashed at the top and bottom.

The fact that Earth is not a perfect globe is a small, important detail. Throughout this book, we'll come across many small differences that are important. All the small things add up, and this makes the weather quite complicated. But if we tackle each of the small things, one at a time, then we will be able to understand how they work together to create weather!

There are lots of examples in everyday life of the centrifugal force. Have you ever been to a céilí dance? At a céilí dance, we sometimes take our partner's hands and we swing them around like spinning tops, faster and faster. What would happen if you let go? You'd fall backwards! That's because when an object – in this case, the dancers – is spinning, there is a force that acts outwards on the object. This force is called the centrifugal force, and this is what pulled the middle of the Earth outwards as it was cooling.

As well as squashing the Earth, the centrifugal force affects many of the weather features we see on our planet, such as hurricanes, tornadoes and low pressure systems. They all have the same type of shape – clouds swirling around a centre and spiralling outwards.

NORTHERN HEMISPHERE

SOUTHERN HEMISPHERE

EQUATOR

23.5°

AXIS

80

70

60

50

40

30

20

10

0

There are two halves to the Earth: the northern hemisphere and the southern hemisphere. The north pole is in the northern hemisphere, and, funnily enough, the south pole is in the southern hemisphere. The two hemispheres are divided by an imaginary line called the **equator**. The equator runs around the centre of the whole globe. Along with the equator in the middle, there are also imaginary circular lines going all the way up to the north and south poles. We call these **lines of latitude**, and they go from 0 (the equator) to 90 (the poles).

Ireland is at 53 degrees north, which is quite far north. It's as far north as Canada and even parts of Alaska – but thankfully it's not as cold here as it is there. There's another line that runs through the centre of the globe from one pole to the other. We call this line the **axis**. If you've ever seen a toy globe, it spins around an axis just like the real thing. You'll also notice that it is tilted. This isn't to make it easier to reach the countries! The real Earth is also tilted on its side, at an angle of 23.5 degrees. This means different parts of the world are closer to the sun at different times, and this causes our seasons!

The Earth is the hottest at the equator. This is because, at 0 degrees, this part of the world is pointed almost directly at the sun all year round. The seasons don't change very much here, because this area is never tilted away from the sun. This means that the weather during the summer and winter is quite similar, and the days and nights are the same length.

It is coldest at the poles. That's because of the tilt of the Earth too. During winter in the northern hemisphere, the sun doesn't get to the north pole at all, because the Earth is tilted away from it. The same thing happens in the winter in the southern hemisphere. During deepest, darkest winter in a place called Rovaniemi in Finland, the sun rises at 11 a.m. in the morning and sets just after 1 p.m. in the afternoon. That's just two hours of daylight!

The position, rotation and angle of the Earth mean that it is a perfect place for us to live. So we'd better look after our planet!

EXPERIMENT #2: ROUND EARTH?

There are a million simple experiments you can do to prove that the Earth is round. Here's the easiest: go to the edge and try to walk off it! OK, that was a joke. Here's a slightly less rude one.

1. Go to a beach around sunset and lie down on the sand.
2. Look at the sun as it tips just below the horizon.
3. Then stand up! You'll get to see it go down again.

THE SURFACE

The Earth's surface is made up of lots of different materials. The most important ones are sand, vegetation, ice and water, and they each affect our weather in different ways.

The great deserts, like the Sahara, the Kalahari and the Mojave, are mostly made of sand, and they are very dry and hot because the sun shines here all year round.

Then there's vegetation, which means trees, plants and fields. Ireland has lots of this! The really important areas of vegetation are the great forests, like the Amazon rainforest.

The north pole and the south pole are covered in ice. Gigantic ice sheets and ice caps also cover the Arctic Ocean, Greenland and the Antarctic.

Then there are the oceans! The Atlantic is 'our' ocean. There is also the Pacific Ocean and the Indian Ocean, the Arctic and the Antarctic Oceans too. That's lots and lots of water! In fact, the Earth is made up of more water and ice than anything else.

ALBEDO AND SPECIFIC HEAT CAPACITY

All of these things – sand, vegetation, ice and water – get hot at a different speed, and this is really important in meteorology. We measure how much radiation a surface reflects or absorbs with a scale called **albedo**. How quickly something heats up from that radiation depends on a thing called its **specific heat capacity**.

ABSORBS · REFLECTS

SAND

Have you ever been to the beach on a hot day and found the sand is too hot to walk on? You probably ran down the beach to get to the water, which is cooler. The sand is hot because it has a low albedo – it absorbs a lot of the radiation from the sun. Surfaces like this that are dry, like the deserts, heat up very quickly. Not only do they absorb a lot of the sun's radiation, they also have a low specific heat capacity, which means it doesn't take a lot of energy to make them hot!

WATER

But why is the ocean cooler than the sand? Water is probably the most important (and complicated!) of all the surfaces that cover the Earth. The sea surface has a low albedo and absorbs a lot of the radiation, just like the sand. However, unlike the sand, the sea doesn't get hot. Instead, it absorbs lots of the radiation but only gets warm very slowly.

This is partly because of just how much water there is! Tonnes and tonnes of the stuff, moving around the Earth in currents and tides. But there's another reason too, and that's the high specific heat capacity of water. That sounds very complicated and it is, but it's also very cool – no pun intended! It's all to do with the tiny pieces of hydrogen and oxygen that make up water. We call these tiny pieces **molecules**. The type of bond between the molecules means that the oceans only change temperature very slowly. It takes a long time for the water temperature to catch up with the land. During the summer in Ireland, the Atlantic Ocean gets to about 16 degrees, but even in the depths of winter the coolest it gets is about 10 degrees. This is the main reason why our climate is so temperate – the ocean around us is never really cold. See, I told you this was going to be important!

VEGETATION

Now, imagine you're on that hot sand again. Ouch! You should skip across the beach to get to a cool patch of grass! Vegetation, like grass, also has a low albedo, but it is not dry like sand. Although it absorbs a lot of the sun's radiation, vegetation contains a lot of water, and we know water takes a lot of energy to heat up. This is one of the many reasons why the great forests are so important. They are part of the balance of the heat coming in and going out of the atmosphere.

ICE

The final material that covers the Earth is ice. The ice at the poles has a very high albedo – it absorbs very little heat – and ice that is covered with fresh snow has the highest of all of the surfaces on the Earth. Not only does the sun not shine on the poles in the winter, but even in the summer, about 95% of the radiation from the sun gets bounced right back off it again.

EXPERIMENT #3: ALBEDO

Take a stone from your garden. We'll also need a pot of water.

1. On a sunny day put the stone and a pot of water side by side sitting in the sun.

2. Notice how quickly the stone gets hot, but the pot of water will only change temperature if you leave it in the sun for ages. This is because the specific heat capacity of stone is lower than water.

GLOBAL CIRCULATION

Global circulation is a way of describing how air moves around the world. This is where meteorologists throw all our knowledge of science and physics into the pot and look at the results.

THE AIR

Remember, the layer of gases that surrounds the Earth is called our atmosphere. The atmosphere is held on by a force we call **gravity**. Gravity keeps most of the gases close to Earth. As we move further up through the sky, the layer of gases gets thinner and thinner – making it very cold and very hard to breathe!

Air is made up of lots of gases. Almost 20% of the atmosphere is oxygen, and the other main gas is nitrogen – there's nearly 80% of that. There's also a little bit of water vapour, which is water in its gas form. There's also a tiny amount of carbon dioxide and some other gases. Mix all that together and you get air!

Oxygen and carbon dioxide are two important gases in our atmosphere because humans and animals use them to breathe. We breathe in oxygen and breathe out carbon dioxide. Plants do the opposite – they use up carbon dioxide and produce oxygen. Another reason to look after our vegetation!

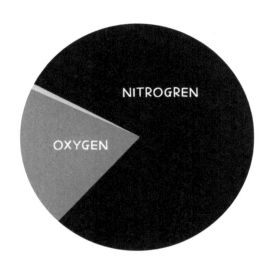

THERMODYNAMICS

So now we know what air is made of – but how does it move? The first thing we need to know about is called **thermodynamics**. This sounds like a strange word, but it makes sense when you break it down – thermo means heat (like thermometer) and dynamics means movement (think of dynamite exploding!). So altogether, thermodynamics is the study of energy. One of the rules of thermodynamics says that energy can't be created or destroyed. It can only be moved around. Heat is a type of energy, so it follows this rule.

Global circulation is about how the sun's heat is moved around and shared. It's not wasted or lost, it's used up and changed into some other form of energy. The heat from the sun moves through the air by a process called **convection**. You'll know about convection if you've ever seen a lava lamp. When the lamp is turned on, the heat from the bulb melts the wax, and it becomes less dense than the liquid it is floating in. The wax begins to rise up through the liquid. When it moves away from its heat source – the light bulb – it starts to cool, so it falls back down again.

Next we can look at how convection works on our planet. There are blobs of air that have the same or similar qualities to each other and we call these **air masses**. The air masses get their qualities – things like humidity and warmth – from the surface over which they form.

Can you guess what an air mass over the Sahara Desert might be like? That's right, warm and dry. And one that forms over the north Atlantic Ocean? Wet and cool! The warm air masses are generally rising – like the blobs of wax – while the cool air masses are falling back towards Earth.

POLAR AIR

TROPICAL AIR

TROPICAL AIR

POLAR AIR

And don't forget, the planet under these air masses is also spinning! With the Earth spinning underneath the atmosphere, a force called the **Coriolis force** comes into play. This force makes the air appear to turn in a circle.

This apparent force, the Coriolis force, is another one of the reasons why weather systems are circular in appearance. There are other reasons (including the centrifugal effect – remember the céilí dancers?) but this is a big one.

Remember the three Cs – the **Centrifugal** force and the **Coriolis** force make **curves** in our weather patterns. Phew – the air on Earth never stands still! This constant movement – rising, falling and spinning – causes the weather.

EXPERMINENT #4: THE CORIOLIS FORCE

You'll need a pin, a large piece of card, a ruler, a pencil and a piece of paper – and a friend!

1. Put the pin in the centre of the card, and cut the paper into a circle. Place the paper on the card so that the pin comes through the centre of the circle.

2. Put the ruler on the paper, pressed against the pin, and run the pencil from one side of the ruler to the other as your friend spins the paper.

3. Although your pencil is moving in a straight line, have a look at the paper when you lift the ruler. See? It's curved! This is due to the Coriolis force.

PRESSURE

The layer of gases that makes up the atmosphere around us is held close to the Earth by gravity. The force that this layer has on the Earth is called pressure.

MEASURING PRESSURE

The pressure of the atmosphere is measured by estimating the weight of the pile of gas from the ground to as high as the edge of the atmosphere, where the gases are so thin we can barely count them. We measure pressure in units called **hectopascals**, or hPa for short.

ALTITUDE

Altitude is a description of how high in the sky something is. As we go higher in to the sky, there is less air around us. The further away from the Earth's surface we move, the less we feel the effect of gravity – the air becomes thinner as we go higher in the sky. And the temperature becomes lower too. At first, the further up in the sky we go the lower the temperature. The air close to the ground is heated by the Earth itself, so the further up we go, the lower the temperature. This changes further up in the atmosphere, as ozone absorbs the radiation of the sun, so it gets warmer there for a while. Then it gets colder again!

Near the surface of the Earth, the temperature generally falls at a rate of about 10 degrees per thousand metres. The highest mountain in Ireland is Carrauntoohil in Kerry, and at its highest point it is 1,038 metres. So if you are standing at the bottom of the mountain and it is 15 degrees, when you get to the top of the mountain, it should be just 5 degrees! Be careful though – remember that there is less air up there too, and that can make it difficult to breathe. People get tired very quickly as they climb mountains. That's why some mountain climbers carry oxygen tanks on their backs.

EQUALISATION

While you're up on that mountain, you might notice that something else happens – your ears make a popping sound! This happens because the air inside your ears is at a higher pressure than the air outside. The popping sound is the extra air escaping, and this helps to equalise the pressure in your ears.

Aeroplanes fly way up high in the sky at heights of about 10,000 metres. It's really cold outside the aeroplane, and there's less air pressure. That's why we have pressurised air cabins. If you open the door of the aeroplane while it is up in the sky, all of the air that is pressed inside the cabin will try to escape really quickly, sucking all the air out of the plane and into the sky. Don't try this, though – you'll be sucked out too!

This happens because air always wants to even itself out – it wants all the pressure to be the same. The air is moving from an area of high pressure (inside the aeroplane) to an area of low pressure (the rest of the sky). The air is always in a rush!

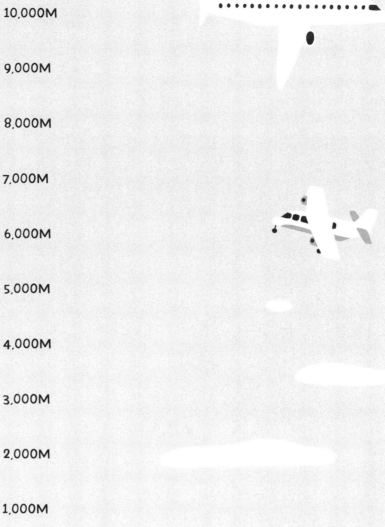

10,000M
9,000M
8,000M
7,000M
6,000M
5,000M
4,000M
3,000M
2,000M
1,000M

EXPERIMENT #5: PRESSURE

1. Take a piece of paper and fold it in two so that it stands on its side.

2. Take a balloon and fill it full of air, until it's almost bursting! The air inside the balloon is under pressure now.

3. Hold the balloon by its mouth and let the air out. Point it at the piece of paper.

The air rushing out of the balloon is the air rushing from an area of high pressure to lower pressure, and it will blow over the piece of paper.

WIND

All of the air rushing from areas of high pressure to areas of low pressure is what makes wind!

ISOBARS

In the weather office, we draw pressure charts with lines called **isobars**. These charts are large maps of the world showing all the areas of high and low pressure. Isobars are lines drawn on the chart that connect all the areas with the same pressure. We can tell which way the wind is blowing by following the direction of the isobars, and we can tell how strong the wind is by how close together the isobars are.

WIND SPEED

The wind is one of the most important things for weather forecasters to predict. We measure the wind in kilometres per hour (km/hr for short), or knots. We also use the **Beaufort Scale** to describe how strong the wind is. This scale is named after the Irishman who invented it, Francis Beaufort.

He created it so that mariners could observe and report on the weather in a way that everyone was able to understand. Before then, people might have had a different idea of what a 'gentle' breeze was, or what a 'strong' wind felt like. Now we can all describe the wind speed in the same way.

1 Light air
1-5km/hr

2 Light breeze
6-11km/hr

3 Gentle breeze
12-19km/hr

4 Moderate breeze
20-28km/hr

5 Fresh breeze
29-38km/hr

6 Strong breeze
39-49km/hr

7 Near Gale
50-61km/hr

8 Gale
62-74km/hr

9 Strong gale
75-88km/hr

10 Storm
89-102km/hr

11 Violent storm
103-117km/hr

12 Hurricane
≥118km/hr

WIND DIRECTION

Along with the wind strength, we also measure the direction. We use eight compass points to describe the direction of the wind: north, south, east, west, northwest, northeast, southeast and southwest. We name wind based on where it's coming from, not where it's going.

When the wind changes direction, we have words to describe that too. 'Back' means to go anticlockwise around the points on a compass. 'Veer' means to go clockwise. So, for example, the wind will back from south to east, or it will veer from west to north.

Here in Ireland, the **prevailing wind** – that's the wind that we get the most often – is the southwesterly wind. It carries mild and damp air up over the country to give us our mild climate. Around the world, there are famous winds that bring special types of weather that are unique to their specific area. In France, there's a wind called the mistral, which is a cold, dry northwesterly wind that blows down through the Rhône Valley and can reach speeds of up to 90 kilometres per hour!

A shamal is a strong northwesterly wind that blows over Iraq and the Persian Gulf. It picks up sand from the desert and causes large sandstorms that can be very dangerous. The chinook in Canada is what we call a föhn wind, and it blows down off the Rocky Mountains. Föhn winds happen anywhere there are tall mountain ranges and bring warm and dry breezes. They even occur here in Ireland. On the eastern side of the MacGillycuddy Reeks, the temperature can be several degrees warmer than on the western side.

In the sky, there are special types of winds called thermals that spin in the sky in the one spot. Birds use these thermals to hover above the world. Plants also use the wind to move their seeds around, sending their pollen flying on the wind. Achoo!

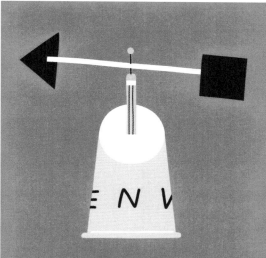

EXPERIMENT #6: MAKE A WIND VANE

1. Stick a cup down onto something heavy. Using a compass, mark north, south, east and west on the paper cup.

2. Cut two pieces of card, one shaped like a triangle and one like a square. Attach them to a straw to create an arrow.

3. Stick a pencil into the cup, and use a pin to attach the arrow to the pencil top.

The point of the arrow will turn into the wind, and you can see where it is blowing from!

Wind has been used by humans down through the ages. Wind carries sailing ships around the world on the oceans and helps hot air balloons to soar above the world. Windmills capture energy from the wind in their sails, and the spinning wheels connected to them have been used for things like milling grain. Nowadays there are wind turbines all over the world being used to harness the energy of the wind to create environmentally friendly power.

CLOUDS

Water can be a solid, a liquid and a gas all at the same time! Remarkable stuff, water. When it's a solid, we call it ice; when it's liquid, we call it water; and when it's a gas, we call it water vapour. There is always a small bit of water vapour in the air around us – even if we can't see it!

CONDENSATION

Have you ever seen your breath on a cold day? Your breath contains a lot of water vapour from your body, and when it hits the cold air, it **condenses**. That's how clouds form too! When the air changes in temperature or pressure, the water vapour starts to condense.

The water vapour in the air changes from a gas to tiny droplets, and we can see it as clouds. Clouds float because they are part of the air itself, but when the tiny water droplets in the cloud get too heavy, they'll start to fall out of the sky – that's rain!

23000 FT

CIRRUS

Cirrus clouds are found high up in the atmosphere at about 30,000 ft. They can look like feathers floating way up high in the sky. These clouds are so high they are mostly made of ice particles and have a shiny and smooth appearance.

ALTOCUMULUS

Below cirrus clouds are the altocumulus, and they can be found anywhere between 6500 ft and 20,000 ft. They look like a sheet of white cotton balls floating on a clear blue lake. Altostratus clouds, which look like a flat, thin layer, can be at around the same height.

NIMBOSTRATUS

Nimbostratus are rain-bearing clouds. They are dark and threatening, and they look like a big grey sheet covering the sky. Watch out above!

6500 FT

There are lots of different types of cloud, and they can be found at different heights in the sky. In Ireland, we usually use metres, but the locations of clouds are measured in feet because that's what pilots use when flying their aircrafts. We use Latin words to describe the different types of cloud: alto means 'high', cirrus means 'curl', cumulus means 'heap', nimbus means 'rain' and stratus means 'layer'.

CUMULONIMBUS

Cumulonimbus are big, heavy shower clouds that can bring hail and thunder. These large towering clouds are formed when there is a big difference between the temperature at the surface of the Earth and the temperature of the air high up. Forecasters call this type of air unstable. We can spot cumulonimbus clouds on a satellite picture because they look bright white and have bumpy tops. They can produce large amounts of rain and hail in a very short space of time.

CUMULUS

Cumulus clouds are formed when the air is moving up and down, warming and cooling, and they can be found at any height. They look like a heap of fluff!

PRECIPITATION

Precipitation is the name we give to anything that falls from the clouds in the sky, which can be rain, sleet or snow.

RAIN

Rain occurs when water in the clouds condenses into raindrops. This happens when water vapour – which is a gas – attaches to teeny tiny particles that float in the air. We call these condensation nuclei. At first the raindrop is too small to fall out of the sky, but it soon attaches to other tiny raindrops nearby, and they come together to form bigger raindrops. When the raindrop is heavy enough, it falls from the sky as rain. If the raindrops are very fine when they come to the ground we call it drizzle.

SNOW

If it is very cold, the rain can become solid and form ice crystals. These are sticky and clump together to form **snow**. Sometimes the snowflakes turn back into rain as they get closer to the Earth and warmer air, but sometimes it's cold enough for the snow to stay as snow all the way to the ground. In Ireland, because we have such a temperate climate, the snow doesn't often reach the ground, or if it does, it melts soon after falling. When it's cold enough, we can get snow on the hills and mountains. **Sleet** is a mixture of snow and rain. We get sleet when it's too cold for rain, but not cold enough for snow!

HAIL

There's also **hail**! Now hail is white and shiny, but it's not the same thing as snow. For a start, we can get hail at any time of the year, even on the hottest day of the year! Hail comes from large cumulonimbus clouds, which can reach really high into the atmosphere where it's very cold. On a hot summer's day, the air at the Earth's surface can get very hot. The heat from the surface rises through convection like big bubbles of warm air.

MIST AND FOG

As it does, it condenses and raindrops form in the usual way. But something different happens in a cumulonimbus cloud. Instead of the raindrop getting heavy and falling to the ground, the raindrop gets sucked up into the atmosphere by the rising air. In the upper atmosphere, it's very cold and the raindrop will freeze. Now it starts to fall down and gets caught up in the downdraught in the cloud. That's the air coming back down – what goes up, must come down! The cold blast carries the raindrop, now a frozen droplet, down through the cloud, where it gets tossed around and smoothed into a ball. It sometimes can get carried back up through the cloud again, tossing and turning.

If the cumulonimbus cloud is very large and powerful, it can carry the raindrop up and down so many times that a large ball of frozen water forms. When the cloud has finished tossing the hailstone around, it shoots it out of the cloud on the downdraught and it can hit the ground hard. Hailstones in Ireland are usually only a millimetre or two across, but in some countries, they can grow to the size of a tennis ball! If you are out and about, watch out for a cold wind that comes suddenly down from above. Once you feel that downdraught, you know it's time to take cover – you don't want to get caught in a hail shower!

Mist and **fog** don't fall from the clouds, but they are also made from water vapour. They form when the water vapour in the air condenses near the ground. It can happen a lot in Ireland around autumn when it gets cold in the early evening as the heat is radiated away. The air just above the ground can hold a certain amount of water vapour as a gas, but when the air cools suddenly, it can't hold as much gas and the water vapour condenses out. Then it just hangs around, suspended in the air. Spooky!

THUNDERSTORMS

Thunder and lightning are a very dramatic type of weather that can seem scary, but when you understand what's going on, you'll be looking forward to the next storm!

STATIC ELECTRICITY

Thunder and lightning come from cumulonimbus clouds. Sometimes, when a cumulonimbus cloud is very high, it goes right over the top of the troposphere and up into the stratosphere. In the stratosphere, it is very cold and all the water vapour in the cloud turns to ice and spreads out. The top of the cloud starts to look like an anvil. The higher and colder the tops of the clouds get, the more likely they are to make lightning.

Lightning is formed when the air inside the cloud moves up and down, creating lots of energy. The air is heated from the ground and then rises through convection. As the air rises up through the cloud it cools and starts to fall again. The water vapour in the air condenses into liquid as it cools, and then it freezes into ice when the temperature gets low enough.

Up and down the air and water vapour and ice go, all bumping off each other. Some of the molecules in the cloud have a positive electrical charge, and others have a negative electrical charge. When they bang off each other, they cause the cloud to become charged with **static electricity**. When the charge inside the cloud becomes too much, it needs to get out! This is when lightning strikes!

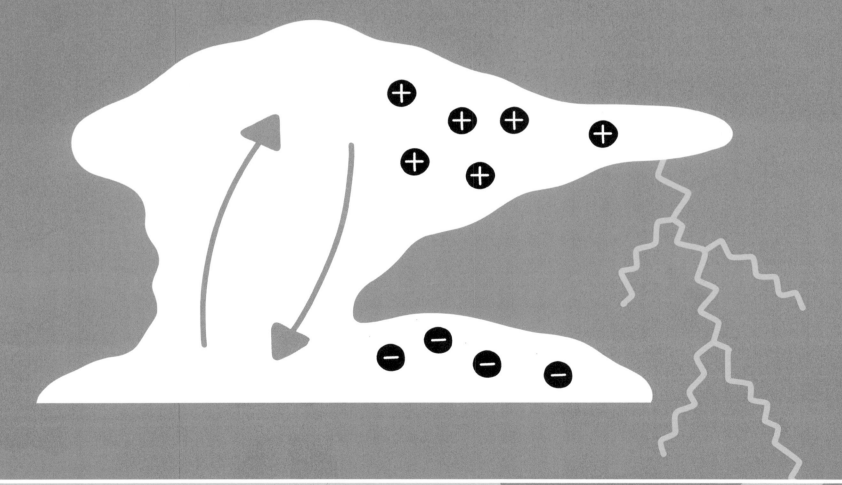

EXPERIMENT #7: STATIC ELECTRICITY

1. Blow up a balloon and rub it against your hair. This will charge it with static electricity. Your hair will stand on end!
2. Now, hold the balloon close to an empty can – but don't touch it! The can should be pushed away from the balloon.
3. The force that is moving the can is the same static electricity that causes lightning bolts.

THUNDER AND LIGHTNING

The **lightning** leaps out of the cloud and gets to the ground as fast as it can, usually by hitting the highest object it can find, like a tree or a tower. Some tall buildings have a metal pole attached at the top to take the charge safely to the ground. Metal makes an easy path for electricity to go through, so a bolt of lightning will hit this rod instead of a person. Thunderstorms can be very dangerous to people, because the lightning strikes can cause fires or explosions.

The sound that comes with the lightning is called **thunder**. Thunder is the sound that the air around the lightning bolt makes as it rapidly expands due to the heat. It can sound scary, but it's exactly like a balloon popping! Because the speed of sound is slower than the speed of light, we often see the lightning before we hear the thunder.

If you count the seconds between the lightning and the thunder, you can work out if the thundercloud is coming closer, or moving further away. More seconds mean it's getting further away. If the lightning and thunder come at the same time, the cloud is right over your head!

Thunderstorms can happen at any time of the year, but it happens most frequently in hot countries during the summertime when the Earth can get very hot. It happens all the time in the equatorial regions, but it doesn't happen all that much in Ireland because of our temperate climate. On the Tiwi Islands in Australia where the weather is very predictable, a big cloud causes thunderstorms nearly every day. Because it's so familiar and comes around every day, local people have given this cloud a name – they call it Hector the Convector!

RAINBOWS

Rainbows are a very special and welcome feature of Irish weather in particular.
To see a rainbow, you need both water droplets and sun, and in Ireland we frequently
get sunshine and showers at the same time.

LIGHT WAVES

But how do rainbows work? Well, the first thing you need to know is that the light from the sun travels very straight and fast – so fast we can't see it moving! When this light hits an object, it slows down, because it's harder to push through an object than air. When it slows, it bends a little bit. Try putting a pencil into a glass of water. It looks bent! This is because the light slows down when it hits the water, tricking our eyes into seeing something that isn't real.

The second thing to know is that the light that comes from the sun is white. But white light is a mixture of seven different colours – red, orange, yellow, green, blue, indigo and violet. This light travels in waves, and each of the different colours of light has its own distinct **wavelength**. The colour with the longest wavelength is red, and the colour with the shortest is violet.

When the rays of light come down from the sun, sometimes they hit raindrops along the way, causing them to bend. As the light passes through a raindrop, the different wavelength of the individual colours means that each colour makes it to the other side at a slightly different angle. This causes the white light to split up into its seven separate rays, creating a rainbow in the sky.

OPTICAL ILLUSIONS

A rainbow is an **optical illusion** caused by sunlight shining through the individual water drops in the rain. An optical illusion is an image that our eyes and brain tell us we can see – but there is nothing physical there to touch. This is why, sadly, you can never reach the end of a rainbow, no matter how close you get!

SHAPE AND COLOUR

Rainbows might look like they stop when they touch the ground, but they are actually shaped like a circle and have no end! We can't see the full thing from where we're standing on Earth, because the horizon gets in the way. Sometimes, people flying high up in aeroplanes can see the full circle of a rainbow.

Some people remember the colours of the rainbows by using this phrase: Richard Of York Gave Battle In Vain – **red**, **orange**, **yellow**, **green**, **blue**, **indigo** and **violet**. Some people remember the colours by thinking of an imaginary person called Roy G. Biv! You can make up your own phrase to remember the colours.

You can even make your own rainbow without showers by using a garden hose. Stand with your back to the sun, spray the hose, and the rainbow will magically appear in front of you!

EXPERIMENT #8: WHITE LIGHT

1. Take a piece of white card and colour it with the colours of the rainbow.

2. Put a small hole through the middle of the card. Now put a pencil through the hole.

3. Spin the pencil very fast. You'll see that the colours blend together and the card looks white!

WEATHER SYSTEMS

The weather around the world can be split up into areas of low pressure, which we also call depressions, and areas of high pressure, called anticyclones.

ANTICYCLONES

We call areas of high pressure **anticyclones**. The air is descending from high up in the atmosphere and moving towards the ground. As the air moves from a high to low altitude it changes. It gets warmer, drier, and clears away the clouds. As the air piles up in the centre, the anticyclone gets stronger and stronger. We usually associate high pressure with fair weather, no rain and blue skies. A good way to remember this is 'Auntie goes cycling in fine weather!' The winds around an anticyclone move in a clockwise direction and typically tend to be light and changing, especially in the centre. Further away from the centre, the winds can get strong. In fact, the winds can get very strong at the outside of the anticyclone. Anticyclones tend to be big and slow – they sit in one position and rarely budge. They are so stubborn that we give them names. One that lives very near to Ireland is called the Azores High, because it lives over a place called the Azores, west of Portugal. In the summertime, if the Azores High takes a little wander and comes to visit Ireland, we can get very nice weather.

Warm southerly winds will become light and variable, and, after a few cloudy days, the clouds will soon disappear and the sun will shine. If the anticyclone stays around for a while, we might call it a **blocking high**. A blocking high will block out the weather fronts that try to move through the air. The weather fronts bring cloud and rain, so when a blocking high comes to stay, it can keep dry and sunny weather over the country for long periods at a time – in the summer. But the same thing can happen in the winter!

There's another area of high pressure that lives near to Ireland – the Scandinavian High. In the media, during the winter, the easterly wind we get in Ireland from the Scandinavian High is called the Beast from the East! Not such a welcome treat! It could bring dry and sunny weather, and will also block out the advancing front, but without our mild southwest weather coming to keep us warm, it can get very, very cold in Ireland. Bring back our southwesterly winds – we'll take the rain, just please turn the heat back up!

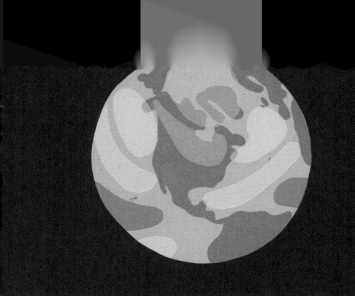

The world is heated unevenly. Some places get too hot – the tropics – and some places get too cold – the poles. Because of this, we get different areas of air with different properties. Some areas have high temperatures and high humidity and some areas have low temperature and low humidity. We call these different areas **air masses**.

Pressure changes in patterns around the world. There is what we call a standard pressure though – it's 1013 hPa. But it's not really right to say anything below that is low pressure and anything above it is high pressure. Because like all things in life, it's all relative! So we distinguish high pressure systems and low pressure systems not by the hPa number, but by the shape and characteristics of the system.

DEPRESSIONS

Now once the air has got from an area of high pressure to an area of low pressure, where can it go? Well, it can't go down into the ground, because it is a solid, so instead it goes up – through the atmosphere. This means that in an area of low pressure, the air is always rising. Usually, we call this a **depression**. From above, it looks like a spiral of cloud.

As the air rises, it forms clouds, and from the clouds we get rain. So if anticyclones bring dry and sunny weather, depressions bring wet and cloudy weather. Low pressure systems tend to be small, fast and much more energetic than their lazy big brother, the anticyclone. Wind moves anticlockwise around a depression and can blow quite strongly.

When a depression gets powerful, we call it a storm. When the storm originates over the tropics, it's called a tropical storm, and if it becomes very powerful, we call it a hurricane! Here in Ireland we don't get too many hurricanes. But we do get regular winter storms. Winter storms and tropical storms are created differently and have different shapes, but they both bring strong winds and rain!

WEATHER FRONTS

Weather fronts occur where one air mass meets another. That's where the real weather happens!

HISTORY

They got the name 'fronts' during the First World War. In the war, lines were marked on the battlefields across Europe to show where the warring armies were. Where the armies met, a line was drawn and it was called the front. During the same time, the study of meteorology and weather forecasting became much more developed and scientists working in the area borrowed the word to describe the place where one air mass comes up against another.

Let's look at a typical Irish weather scenario. There is high pressure over the country, and the day starts out fairly ordinary with only a little bit of cloud. But here comes a **warm front**! Out in the Atlantic, a new air mass is moving towards us. It is warmer than the air that is over Ireland at the moment.

As this warm air meets the air that is already here, it rises up and over the top of it. The air at the ground is still the same as it was, but in the sky above us, we can see that a thin layer of cloud has started to form. So for us here on the ground, we see the day going from bright and sunny to hazy, then the clouds get whiter and thicker until they become low and dark and it starts to rain.

Once the warm front goes through, here in Ireland we often find ourselves in what we call a warm sector. In a warm sector, it can usually feel quite humid, or often cloudy with a little drizzle. But in the summertime, it can become warm and sunny. If we're lucky enough to find ourselves with a blocking high nearby, this warm and sunny weather can last for days or even weeks. This doesn't happen very often though.

WEATHER CHARTS

On a weather chart, a warm front is drawn as a red line with a half circle. A blue line with triangles is used for a cold front. A purple line with both circles and triangles is used for an occluded front. When you look at a weather chart, you can tell which way the front will go by looking at which way the wind is coming from. We know that the wind goes clockwise around an anticyclone and anticlockwise around a depression, so we can work out which way the weather front will go. The wind will always blow in the same direction the front is travelling.

The most common way to see fronts crossing Ireland is from west to east, but sometimes they come from the south, and, although rarer, from the east or north too. Mostly, fronts pass by quickly, usually within a matter of a few hours. We're lucky – in some parts of the world fronts can hang around for days or even weeks!

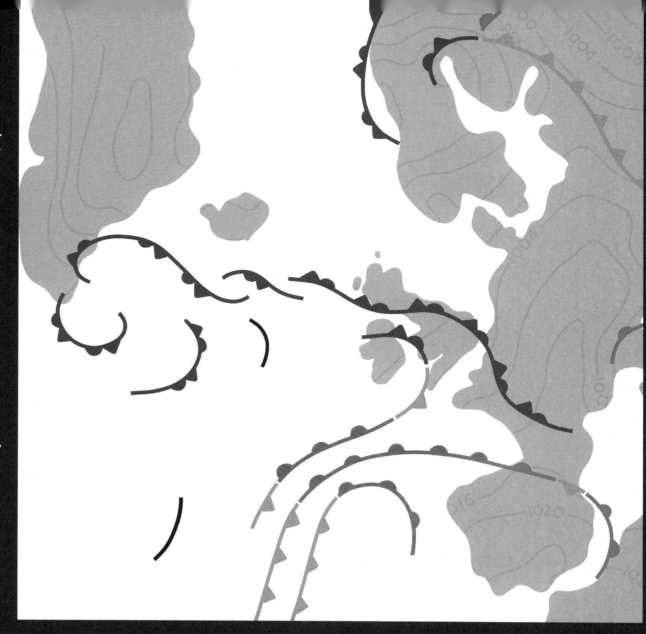

Usually, right after the warm front goes through, a **cold front** heads straight on behind it! With a cold front, the weather changes from a higher to a lower temperature, so the air that comes in is colder. Cold air sinks under the warm air, pushing it up and out of the way.

When air rises, it forms clouds, just like a warm front, but with a cold front, the clouds tend to bunch together and form much more quickly. When a cold front goes through, the skies over Ireland tend to first brighten, with sunshine and showers following behind. The rain marks the change to colder air.

When a cold front catches up with a warm front, the transition from one air mass type to another becomes less clear. We call this area an **occluded front**. Occluded fronts also tend to bring rain and they wrap back around the low pressure that is driving the fronts through.

CLIMATE

Climate is the word we use for all of the weather conditions all mixed up together. It includes the patterns made up by pressure, temperature, precipitation and wind.

The weather changes from season to season and from day to day – some people say it changes every hour in Ireland! But the climate changes only over millennia. Here in Ireland, we can get snow days and heat waves, but we would still describe the climate as temperate. On earth, there are hot, wet, cold and dry climates – and everything in between!

THE GLOBAL CLIMATE

We divide the earth into regions with imaginary lines of latitude. Some of these lines are given special names to help us describe the climate we can expect there. These are called the Arctic Circle, the Tropic of Cancer, the Tropic of Capricorn and the Antarctic Circle.

 The north pole is at the very top of the world, at a latitude of 90 degrees north. The Arctic Circle, at 66 degrees north, is very cold and dry.

 The Tropic of Cancer is hot and dry at 23 degrees north. Ireland lies in between, at 53 degrees north. In this area, the climate is generally not too hot and not too cold – we call this **temperate**.

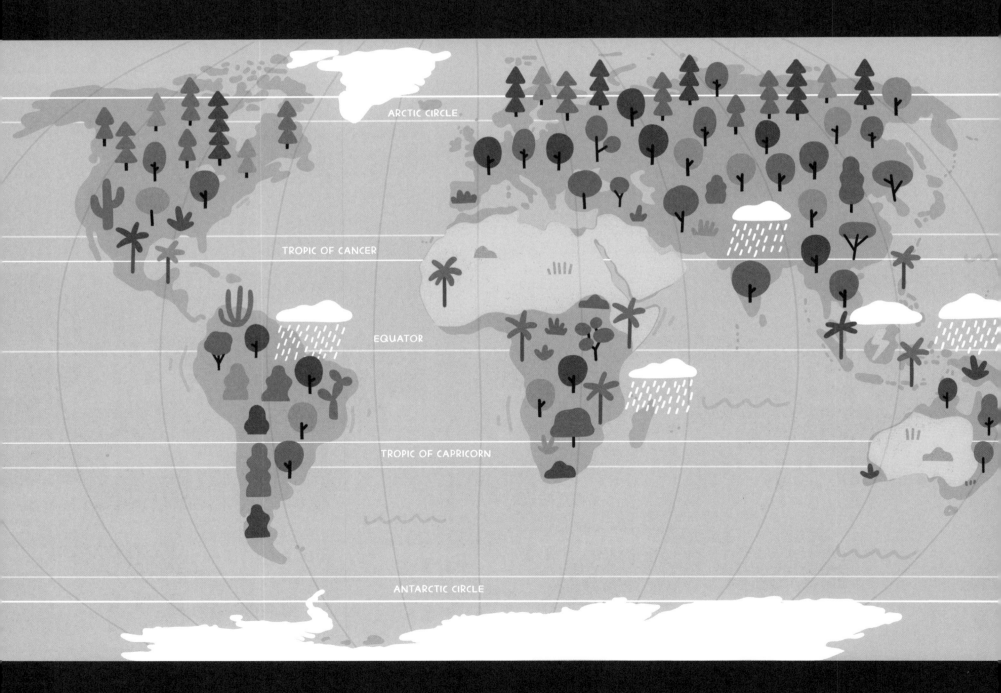

ARCTIC CIRCLE

TROPIC OF CANCER

EQUATOR

TROPIC OF CAPRICORN

ANTARCTIC CIRCLE

 The Tropic of Capricorn lies at 23 degrees south and the Antarctic Circle lies at 66 degrees south. Between the Tropic of Capricorn and the Antarctic Circle, the climate is also temperate.

 Below the Antarctic Circle, the climate is cold and dry. The south pole is all the way down at 90 degrees south.

CLIMATE TYPES

In between the two Tropics, it's warm and wet. This is called a **tropical** climate. In the equatorial regions of Africa and South America where there are large rainforests, it rains a lot of the time. The area of the world between China and Australia gets a lot of rain too, for example in the Philippines and Indonesia. Sometimes there are seasonal rains called **monsoons** that are very heavy. It might rain non-stop for a month! We don't get monsoons in Ireland, although sometimes in July it can feel like we do!

There are other regions of the world where hardly any rain falls at all. These areas are called **deserts**. Not all deserts are hot and dry – some can be cold, like the Antarctic and the Gobi. The thing all deserts have in common is that they only get a tiny amount of rain in a year.

In other parts of the world, the weather can vary, much like in Ireland. But in places like Australia, the United States and right across mainland Europe into Asia, the weather can be much more predictable. That's because in these places the weather is linked to the land. In the summer, the continents get warm as the sun shines and steadily heats the air. Anticyclones build up, and there might be very little change in the weather for weeks on end as they become blocking highs and stop the movements of fronts. Then, in the winter, the heat radiates out of the earth. Again, anticyclones build up and it can be dry and clear – and cold! The change between the seasons can be a dramatic transition with autumn and spring often arriving in an afternoon, as temperatures rise or fall overnight.

The hottest places on earth are in the deserts. Death Valley in California in the United States is thought to be the hottest place on Earth with temperatures regularly reaching 47 degrees Celsius, and a high of 56.7 degrees was recorded in the summer of 1913. A combination of the desert and the terrain around it is what causes the heat to be trapped and held there. Temperatures in Al Aziziyah in Libya regularly reach highs of 48 degrees. A place called Dallol in Ethiopia has the highest average temperature at 35 degrees – that's the temperature averaged out between day and night.

The coldest place on Earth is at the south pole in Antarctica, where temperatures as low as -93 degrees are recorded. But nobody lives there, except for some scientists who live on special testing stations for a few weeks at a time. The coldest inhabited places on Earth are in Russia, where temperatures go as low as -45 degrees in winter – now that's cold! And the children still go to school! In fact, school is only cancelled when temperatures go below -52 degrees. Brrr!

The wettest regions are around the tropics and this is where we find the equatorial rainforests. Here, plants and trees grow to enormous sizes. They grow so big because they have lots of water and warm soil to grow in. In Ireland we need greenhouses and watering cans! But the wettest place of all is in India and that's because of the monsoon rains. A place called Mawsynram has an average annual rainfall of nearly 12,000 mm.

THE CLIMATE OF IRELAND

Remember the story of Goldilocks and the Three Bears? Ireland has what I like to call a Goldilocks climate – just like the porridge, it is neither too hot nor too cold.

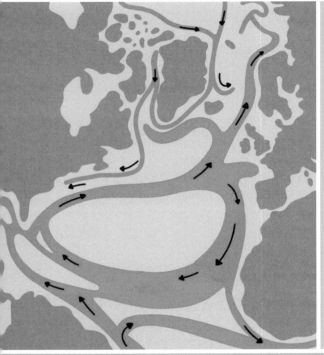

THE GULF STREAM

In Ireland, the surrounding oceans keep us relatively warm all year round, which is largely because of the **Gulf Stream**. The Gulf Stream is a current of warm water that runs through the north Atlantic Ocean, beginning in the Gulf of Mexico and stretching up along the eastern coast of the United States, where it heads eastwards and splits in two. One branch heads down towards Africa and is called the Canary Current, the other – the North Atlantic Drift – takes a path towards northern Europe, passing by us here in Ireland.

The Gulf Stream keeps the water around our shores between about 7 degrees at its coldest in March to about 15 degrees at its warmest in August. As well as being fast moving, this ocean current also contributes to our temperate climate. The air above the warm current holds onto its temperature and doesn't cool as it moves north away from the equator. Our prevailing southwesterly winds come from the Atlantic and bring this mild air to us. Despite the fact that we are quite far north – as far north as some parts of Canada and Russia! – we have quite a mild climate due to the Gulf Stream.

PRECIPITATION

Our wettest months are from October to January, and each month we might see about 130 mm of rain. The most rain in Ireland falls over the western half of the country, which is partly because of the relatively warm Atlantic Ocean. The shower clouds coming off the Atlantic are formed by air rising over the warm ocean underneath. They drop their rain on the west coast of Ireland, but as they move over the cooler land, they lose their energy and fizzle out. Listen out to the forecast and you'll hear 'showers will become confined to western coasts'!

Most of the rain falls on the hills and mountains in the west. As the air rises up over the hills, it condenses further into clouds and makes even more rain. So if you like the rain, then the best place to live in Ireland is on a hill in the west! The driest area in Ireland is the east. Dublin and the northeast coast have the least amount of rain. This is partly because these areas are sheltered by the Wicklow Mountains – they act like a giant umbrella! The prevailing southwesterly winds sweep up over the country, carrying the rain-bearing clouds.

When the clouds reach the mountains, they are forced to rise up further into the atmosphere to get over the ranges. This causes the air to condense and spill out the rain, just like the westerly hills. All the rain is gone before the clouds make it across the country to Dublin! The wettest day in Ireland so far was in Cloone Lake in Co. Kerry on 18 September 1983, when 243.5 mm fell in one day. That's more than the average monthly rain for even the rainiest month – and it fell in one day!

WIND

The winds in Ireland are mostly south westerly winds. It can get very stormy in the wintertime when Atlantic storms come along the west coast. The strongest winds ever recorded in Ireland came from a hurricane! When Hurricane Ophelia hit Ireland in 2017, a wind gust of 191 km/hr was measured at Fastnet Rock off the coast of Cork. Avoid the west coast if you don't want to live in a windy place!

TEMPERATURE

The average temperature in Ireland is around 9 or 10 degrees. This is the temperature taken as an average of day and night and spread over the year. July is the warmest month and January is the coldest. The warmest it gets is usually between 18 and 20 degrees, which is not terribly warm. The warmest it ever got was 33.3 degrees in Kilkenny in 1887.

The coldest temperature ever recorded in Ireland was in Markree in Sligo in January 1881 and it was -19.1 degrees. The highest temperatures are inland and so are the lowest. That's because the warm sea that surrounds us keeps the temperature stable at the coasts. If you like it hot, then the best place to live is the midlands, but if you don't like the cold, then the best place is at the beach!

The sunniest place to live in Ireland is the southeast of the country. That's because it's closest to the heat of Europe. Also, the showers that develop over the Atlantic Ocean to the west have often run out of energy by the time they get to the east coast of the country. If you like the sun, try the southeast coast, but if you prefer the rain, head west!

PALEOCLIMATOLOGY

The word paleoclimatology (which sounds very serious!) relates to the climate that happened over the millennia since the world began. The word paleo means ancient, and climate means – well, you know that one already!

ICE AGES

The world did not always look as it does now. There was a time when the ice that we see at the north and south poles extended through the Tropics of Cancer and Capricorn. In fact, the whole of Ireland and Europe were covered in ice! Scientists call this an ice age. During an ice age, the temperature of the earth is much lower than it is now.

There is evidence here in Ireland of the movement of **glaciers**. Glaciers are large sheets of compacted ice and snow that move very slowly. As they move, they drop pieces of rock and earth that have been trapped in the ice. Scientists can examine rocks and the shape of the hills and mountains around us to build a picture of how the world looked before we were here.

The landscape in the northwest of Ireland is a great place to see examples of glacial movement. Benbulben in Co. Sligo has its distinctive shape due to the formation and movement of glaciers.

Scientists can also drill big holes in the north and south poles and pull out enormous sections of the ice. When we look at these ice cores, we can see that they are made up of layers. This ice formed over millennia as snow fell and froze on top of previous layers of snow and ice. We use this to see how the climate over the poles has changed through the years.

There have been several ice ages in the past, and in between, there are periods of warming. We call these times interglacial periods. We are in an interglacial period right now! The last ice age ended 2.6 million years ago. It is suspected that man first appeared at around this time because human remains from 2.8 million years ago have been found.

VOLCANIC ERUPTIONS

We can also tell when huge **volcanic eruptions** took place by looking at ice cores. When volcanoes erupt, they shoot gigantic clouds of ash up into the atmosphere. This spreads out through the air and eventually falls back to earth. The ash settles on the ground, and when it snows again the ash is sealed into the ice for scientists to find later. The bigger volcanic eruptions can put enough ash into the atmosphere to block out some of the sun's radiation. This can cause the earth to enter a cooling period. In 1991, Mount Pinatubo in the Philippines erupted. It was called a cataclysmic eruption because it was so huge – it poured nearly 20 million tons of ash and chemicals into the stratosphere. When it dispersed, it caused the temperature of the world to reduce by half a degree from 1991 to 1993. When we consider the average global temperature between 1951 and 1980 was about 14 degrees – a change of half a degree is quite a lot!

EXTINCTION

Scientists suspect that climate change brought an end to the age of the giant dinosaurs about 66 million years ago. The impact of a massive comet may have thrown clouds of ash and chemicals into the air, causing a cooling period. This led to an **extinction** event, in which many dinosaurs died, while others shrank and evolved to adapt to the new climate. Maybe nature didn't care too much about giant dinosaurs. Let's hope she cares more about us!

There are lots of different events that could explain why the climate of the earth changes so much. Some of these include changes in the earth's atmosphere, the movement of continents and ocean currents, changes in the sun's energy, the eruption of volcanoes, small variations in the distance from the earth to the sun and the slow change in the tilt of the earth's axis. But these past changes were very slow and gradual, and continue to happen naturally. However, climate change caused by human activity is only quite new, and it is happening much faster!

CLIMATE CHANGE AND HUMANS

As the number of people on the planet continues to grow, the amount of food and energy we use increases faster and faster. It has been observed by scientists around the world that the temperature of the earth is rising, and this is caused by the activities of humans. This is due to something called the greenhouse effect.

GREENHOUSE EFFECT

Have you ever walked into a greenhouse? It's hot in there! This is because the glass traps the heat of the sun and keeps the air inside warm. In the atmosphere, the gases in the air behave like a greenhouse. One of those gases is **carbon dioxide**. Remember, when we breathe, we take in oxygen and release carbon dioxide, while vegetation takes in carbon dioxide and puts out oxygen.

When humans cut down forests to make room for crops and animals, that reduces the number of trees taking in carbon dioxide. As well as that, we burn **fossil fuels** – things like oil, coal, gas and peat – to make electricity or to run our cars and aeroplanes. This makes even more carbon dioxide. All this extra carbon dioxide means the atmosphere is trapping more and more heat, warming the earth. Scientists have also discovered that the increasing amount of carbon dioxide makes the oceans warmer. When the oceans get warmer they get higher because warmer water takes up more space than colder water, so the oceans expand and the sea level rises. This can cause the erosion of coastlines and the destruction of natural habitats. If the water continues to rise, whole islands – maybe even parts of Ireland! – could disappear!

ICE CAPS

Another problem is the melting **ice caps**. As temperatures increase, the ice that covers Greenland and Antarctica melts and flows into the ocean. As well as causing the sea levels to rise, it has another effect. This ice is made from fresh water, but the oceans are made from salty water. Mixing the two types of water together causes changes in the way the oceans move. The melting ice caps will also cause the earth to heat further. Ice – and especially snow-covered ice – has a very high albedo and absorbs very little of the sun's radiation. If the area of the earth that is covered in ice is shrinking, more radiation will be absorbed – which means even more heat.

With so many changes going on in our air and in our oceans, it's obvious that there are going to be changes in our climate beyond the weather simply getting warmer. This is why we refer to what's happening as **climate change** rather than global warming.

Scientists are still trying to figure out what exactly climate change will mean to each country. The current thinking is that here in Ireland, extreme weather events will still happen in the same way they do at the moment, but what will change is how frequently they happen and they may become a little more extreme.

Around the world, though, things may be more serious. As droughts become longer, people will struggle to farm for food as they have done in the past. As monsoons become less predictable, last longer or bring more rain than they usually do, flooding in affected areas could displace large numbers of people. If they are shorter or bring less rain, this can disrupt the agriculture of the region, which will also cause people to move.

The **displacement** of people will become a problem as the world struggles to adjust to very large numbers of migrating populations. We need to be aware of the struggles of other people, sometimes from different cultures to ours, who will need to move around the world to find places to live where they can have food and shelter.

Climate change is more than just rising sea levels, whether coastal houses will lose their garden or if a golf course needs a better wall. Climate change is about more than extra rain, or the weather getting too hot or too cold. Climate change will affect the number of plants and animals that will be able to continue to live in their habitats, either in the oceans or on land. Climate change will affect how we live in our own country as well as how we make room for other people to live in our country with us. All these problems mean climate change is perhaps the greatest danger facing the planet today.

37

WHAT CAN WE DO?

If we want to make a difference to how we affect the climate, we need to think about what we are doing each day that affects the world around us. The two big things that humans do are eat and use energy.

FOOD

When it comes to food, we need to try to grow as much of our own food as possible. Transporting food around the world creates environmental damage, so the closer the food source the better. Eating seasonal fruit and vegetables that are grown close to home is not only helping the world's climate, it is also a very healthy way to live!

We can grow lots of fresh and tasty fruits and vegetables here in Ireland – in greenhouses, gardens, on balconies or even windowsills. Try it yourself! We can also try to cut down on the amount of meat we eat. Large areas of land are needed to produce the grass that cattle need to eat, and many of the rainforests are being destroyed to make room.

And then there are the cows themselves! They produce a gas called methane when they digest their food, which is another of the gases that increases the greenhouse effect. Consider having at least one meat-free day a week, and on other days eat locally sourced meat from Irish animals and farmers. Always think about the environment when you choose your food.

RECYCLING

Always recycle your packaging, but better than that, try to use products with as little packaging as possible. Avoid single-use plastics. These are things like plastic cups at the water fountain, straws and little plastic bags.

Bring your own bottle and refill it rather than buying bottles of water all the time. Make your own sandwich wrappers from older materials that you already have, or use a lunchbox rather than wrapping your food in tin foil or plastic wrap.

ENERGY

Energy is another thing that we need to think carefully about. Cars and other forms of transport get their power from fossil fuels, which creates carbon dioxide. Walking or cycling to school or work is much healthier, and it's much better for the environment too. And more fun! If you cannot walk, cycle or take public transport to school or work, you can still try to improve on what you do – you can carpool! Carpooling is when people taking the same journey share a car. Instead of the car going to the school with just one or two children, you take your friend with you. If everyone that can't walk or cycle to school did this, it would halve the number of cars on the road!

What about other types of energy? Always turn off electricity when you're not using it. Turn off lights when you leave the room. Don't just use the remote control to turn off the TV – go right over to the wall and flip that switch. Before you turn on the heat, put an extra jumper on! This will also improve your family's electricity bills as well as saving the environment. Cleaning water to make it safe for us to use takes a lot of energy, so the less we waste, the better. Use a rain butt to collect rain water in your garden if you can. This will reduce the amount of water that needs to be cleaned. Take showers instead of baths, and never leave the tap running while you brush your teeth.

Our modern lives make it very easy to take the easy option, but easy options often harm the planet the most. Humans need to learn to be less lazy. Think about how your actions will affect the rest of the earth whenever you do anything – the planet needs your help!

FORECASTING

Forecasting means using what we know now to make a prediction about the future. Economists forecast what is going to happen with the economy by looking at the market. Salespeople forecast what toys to stock at Christmas based on how popular they are on *the Late Late Toy Show!*

Weather forecasting is very complex because there is so much information and many complicating factors involved. The earth is huge and there are many variables, like temperature, pressure, humidity, the time of day and the month of the year. The more variables that we have to deal with, the more outcomes are possible – and a small change can mess up the whole prediction. For example, a butterfly in Argentina might flap its wings, causing a tiny change in the air, and the end result is a hurricane over Florida! We call this the **butterfly effect**, or chaos theory.

Many people complain about weather forecasters 'never getting it right'. The development of computer models means we are improving all the time, but because of chaos theory, there will always be a limit to how far into the future we can forecast.

We have to know what the weather is like now before we can forecast what it's going to be like later. To do this, we observe the weather and make a record of the conditions.

Around the world, meteorological observers take measurements in the same way at the same time of day. The measurements are all taken at ten minutes before the hour on the clock and then transmitted around the world. That means someone in Australia can take a weather measurement at ten minutes to three in the afternoon, and at three o'clock I can see it in my office in Dublin!

So what type of weather is measured, and how do we record it?

PRECIPITATION

Precipitation is anything that naturally falls from the clouds, most commonly rain. We don't count skydivers or pigeons! We measure rain with a **tipping rain recorder** and use millimetres (mm) to describe how much has fallen. In the past, rainfall was measured by lowering a glass jar into the ground. Every hour, an observer would take the jar out of the ground to see how much rainwater was in it. But this only told us how much rain had fallen, and not how fast it was falling. Now, we use a tipping rainfall recorder, which has a funnel to guide the rain down into a bucket with a tiny see-saw. Each time 0.1 mm of rain falls onto one side of the see-saw, it tips over. As it tips over, it sends a signal to a computer. This is very useful to forecasters because we can measure the rate at which the rain has fallen.

WEATHER

VISIBILITY

Visibility is how far a person can see at a certain time and is measured in metres. Observers at airports estimate this by looking at certain landmarks and recording whether or not they can be seen. Mist and fog cause poor visibility. We call it mist when visibility is reduced because of the water droplets, but when it's reduced below 1,000 metres, we call it fog. Other things that make the visibility poor are haze or pollution, so visibility in big cities is usually poorer than in the countryside. Visibility is important for the safety of motorists and aviators.

PRESSURE

We measure pressure with a **barometer**. In the past, mercury barometers were used, but these days, all our measuring instruments have very sensitive sensors. The barometer uses three small cans that have had all the air sucked out of them. The cans are measured by very fine instruments to work out changes in the amount of pressure around them. We describe the pressure level in hectopascals. The pressure tendency is also recorded, to see what way the pressure is changing.

WIND

The wind is measured by an **anemometer**. It uses two separate devices on one sensor to measure both the wind speed and direction at the same time. An arrow spins and tells us which direction the wind is coming from. The cup on the other arm spins around to tell us how fast the wind is blowing. Wind speed is measured in knots. We convert knots to the metric system of kilometres per hour, but sailors and pilots still use knots! There is also the Beaufort Scale, which describes wind speed in a different way. Force zero – no wind at all – only happens in the eye of a storm or in the centre of an anticyclone. Beaufort force 12 describes hurricane-force winds.

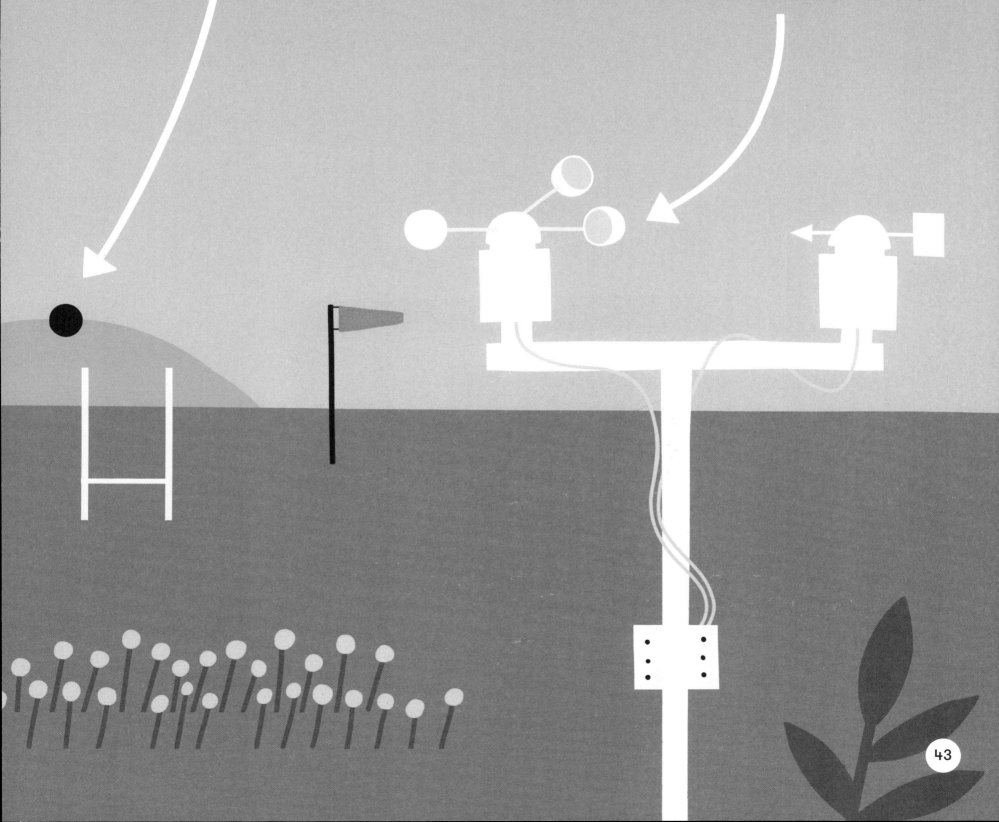

TEMPERATURE

We use a **thermometer** to measure the temperature. In Ireland temperature is measured in degrees using the Celsius scale. We take the temperature of the air at two metres above the ground, and everyone around the world puts their thermometer at the same height so we are all measuring the same way. The air temperature changes in and out of the sun, so we use a something called a Stephenson screen. This box keeps the instruments out of the direct sun and has louvre doors – like little blinds – to let the air in and out. As well as the temperature of the air, we also measure the temperature of the ground and the temperature just beneath the surface of the ground.

DEW POINT

We also measure something called the **dew point** in Celsius. The dew point is the temperature at which the water vapour in the air changes from a gas to a liquid and settles as dew. We see dew in the morning, sitting on the grass, and if the temperature is below zero, the dew can freeze and turn white, covering everything in frost.

RELATIVE HUMIDITY

One of the things we use the dew point to calculate is called the **relative humidity**, which tells us how much moisture is in the air. When it is very humid we feel warmer. When it is dry – less humid – we feel cooler. The reason for this is the way human beings control their body temperature. We release heat by evaporating water off the top layer of our skin – we call this sweating! When water changes from a liquid to a gas, it absorbs energy from the skin, cooling us down. If there is a lot of moisture already in the air, it is harder for more moisture to evaporate into it. So if the relative humidity is high, we feel warmer, and if it is low, we feel colder.

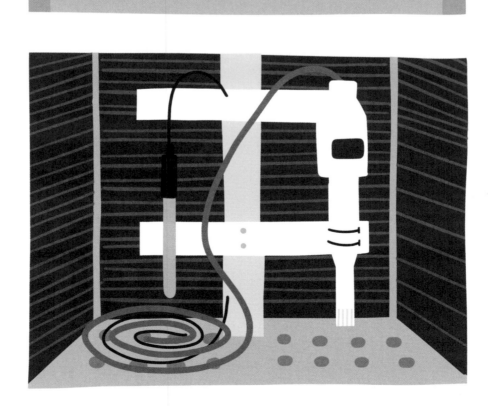

CLOUDS

We also measure the amount and types of cloud. The cloud amount is observed and measured in **okta**. The sky is divided into eighths and it is estimated how many of those eighths have cloud in them. A measurement of zero okta means a clear sky, while eight okta means the sky is completely overcast. The cloud height is measured too because the height of a cloud gives us an idea of what type it is and what type of weather it might bring.

SUNSHINE

We also observe the number of hours of sunshine recorded in a day – there usually aren't very many here in temperate Ireland. Sunshine hours are recorded using sensors these days, but in the past, we used a crystal ball! The crystal ball was balanced in a device called a Campbell Stokes recorder, which was placed facing the sun. A piece of paper ran behind the crystal ball and as the sun shone, the rays were concentrated, burning a little hole. As the sun moved across the sky, the area that burned moved too. At the end of the day, observers counted up the burned paper to calculate the hours of sunlight.

SNOW

Snowfall is measured with a ruler! However, during a blizzard – when snow falls during strong winds – we can only estimate how much has fallen. In a blizzard, the snow is blown by the wind and piles up unevenly against buildings, trees, cars – basically anything that can get in the way!

When we have all the weather observed, we use a computer to plot the information onto a chart. The chart lets forecasters see at a glance what the weather is and helps us predict what it will be like in the future.

SATELLITES

The word satellite refers to anything that revolves around a planet. That can mean another planet or, in this case, a manmade machine. Ever since man first went into space, we have been using satellites to help us understand the earth better and to communicate with each other.

INFRARED

Satellites measure the temperature of the earth by the amount of radiation it is emitting. Computers are then used to change the temperatures into shades of black and white. Low temperatures are given light shades, and high temperatures are given dark shades. This is called the **infrared** (IR) satellite image. Clouds are cold and they appear bright white on an image. The whiter the cloud, the colder it is, which tells us how high it is. When we know how high a cloud is, we can work out which type of cloud it is. We can see the oceans on the IR satellite picture because they are dark and relatively warm compared to the clouds. The land can also be very dark in some places where it is warm. Sometimes it is hard to see where the land and sea meet because the temperatures are very similar.

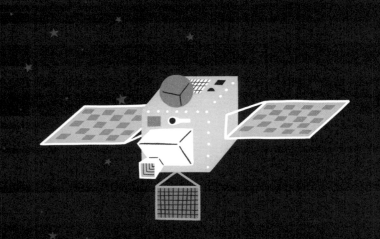

WEATHER SATELLITES

In meteorology, we use **weather satellites**. These take lots of measurements from the earth and use computer code to change the measurements into pictures so we can see what is happening in the atmosphere. There are different types of weather satellites. Some stay in one spot above the earth and are called geostationary satellites. The other type rotates around the poles of the earth, and these are called polar orbiter satellites.

Hurricanes have a very distinctive shape, like a spinning top. Look for hurricanes over near the Caribbean.

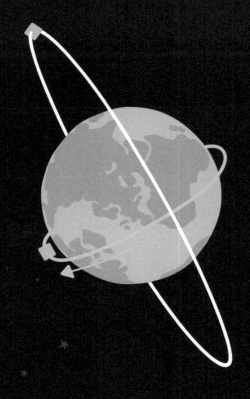

VISIBLE SATELLITE IMAGE

A **visible satellite image** is a picture of the earth from above. The visible satellite image is made by measuring how much of the sun is reflected back to the satellite. Computers then change these measurements into pictures. IR images can be made all day long, but visible satellite images use reflected sunlight, so are only available when the sun is shining. Since visible images record reflected light rather than radiation, they tell us how reflective an object is, but nothing about its temperature. Snow has a very high albedo and reflects almost all the light that shines on it. So on a visible satellite image, snow appears bright white. Water has a very low albedo and absorbs almost all the light that shines on it, and always appears very dark on a visible satellite image. The land's albedo depends on whether it is made from sand or thick vegetation, so it has varying shades. Clouds reflect a lot of light and appear bright on a visible satellite image. The higher the cloud, the more likely it is to have ice (which reflects light even more), so high clouds appear bright white.

WATER VAPOUR IMAGE

We can see the shape of weather systems that are developing in the atmosphere by looking at another type of satellite image – the **water vapour image**. This image looks at the amount of water vapour in the atmosphere and gives forecasters a good idea of the position of the strong winds. These strong winds play a big role in the development of weather systems at the surface.

A warm front is identifiable on a satellite image by the big cirrus shield that sweeps out ahead of it.

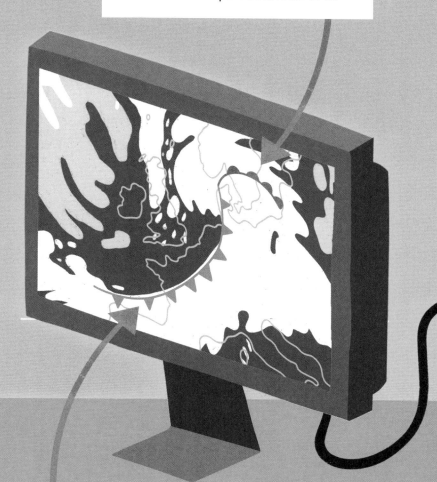

It may look like there is a line of cloud separating Spain from Portugal – nope! That's snow lying on the tops of the Pyrenees. There's snow on the tops of the Alps too.

It's much easier to see a cold front on the surface because there's a clean edge on the back side.

RADAR

Radar stands for Radio Detection And Ranging. A radar is a large machine that sends a radio signal out into the air, which bounces off objects and returns to the machine. This gives forecasters an idea of what is going on in the sky.

RAINFALL

When a radio beam hits a raindrop in the sky, it bounces back. The machine then measures how long each individual beam takes to get back and uses these numbers to create a picture of what the rainfall looks like. These pictures look a bit like a colouring book. Blue is for light rain, yellow is a little heavier, while orange and red are for heavy rain. If you ever see purple on the radar, you better make sure you're indoors!

We need to be careful when looking at the radar, as there are some things that can confuse it. Sometimes the radar image will detect rain that is falling through the sky, but this rain may dissolve before it reaches the ground. This can happen regularly during the summer when the air is warm and dry. The radar can also get confused by drizzle. When it is drizzling, the raindrops are very tiny and the radar might mix up light drizzle with dense cloud. It is important to keep an eye on all the data that a forecaster has to make sure we have the full story at all times.

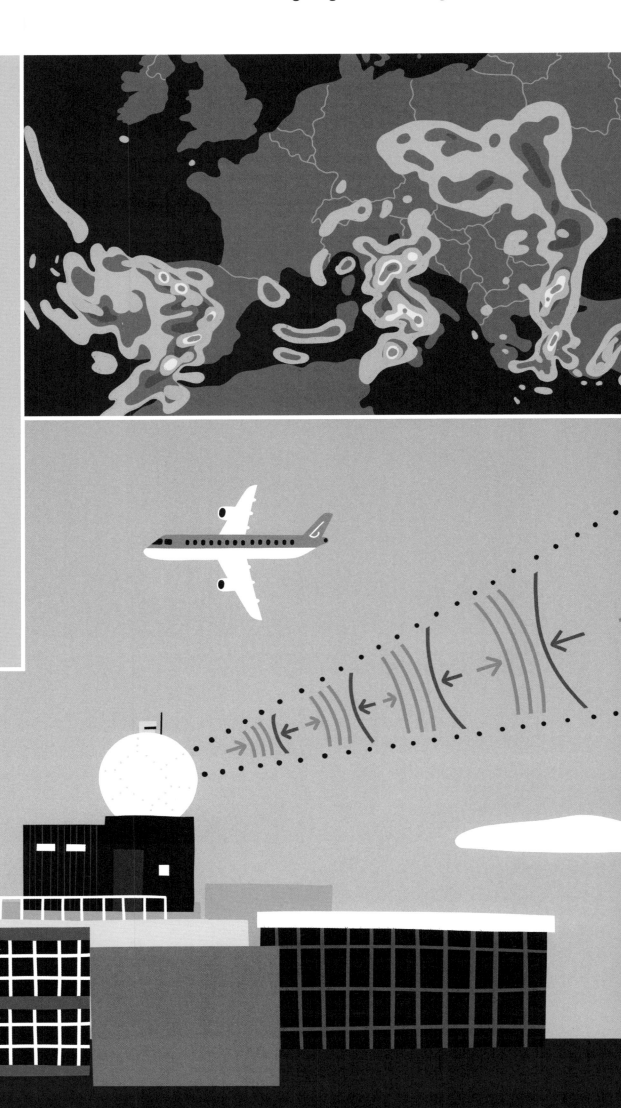

AIRPORTS

Radar imaging is also used to track aircraft or other objects in the sky, so we often find them at airports. Aeroplanes and helicopters are easily identified on a radar image because they are large and fast-moving objects. They have large domes built around them to protect the delicate machinery inside, and they also protect nearby people from the rapidly rotating arms of the radar. The **radome** (radar dome) is made from a thin material that slightly enhances the signals being transmitted from or received back to the radar. The radar itself is inside the radome.

The radar is probably one of the most useful tools for people who need to know whether it's going to rain. In Ireland, we have a rainfall radar at Dublin Airport and at Shannon Airport, and we also use the radar from the British Met Office. We mix the three together so that the whole country is covered.

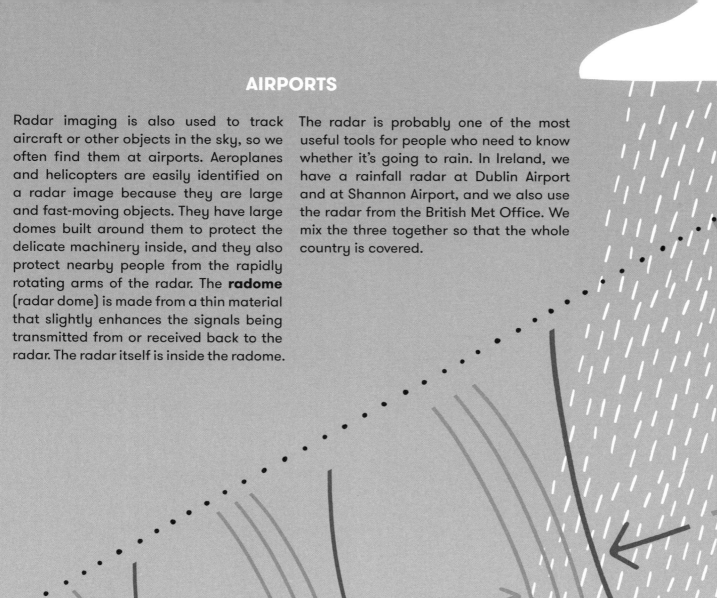

ANAPROP

Sometimes the radar beam can hit off large objects nearby and the returning signal can be mistaken for rain. When the radar is first put in place, computer technicians work to remove these signals from the computer programs. But sometimes the weather conditions can make the radar signals reach further away, mostly in times of high pressure or 'stable' air. When this happens, our radar beam can send back signals from large stationary objects like mountain ranges. Experienced forecasters know to watch out for this feature, which we call **anaprop** (short for anomalous propagation). It will appear as stationary rain in a well-formed shape – and rain rarely does that!

COMPUTER MODELS

The biggest change in weather forecasting in the past 30 years has been the advancement of computer models. Computer models use mathematics to apply the laws of physics to movements in the atmosphere. We call them Numerical Weather Prediction models (NWPs).

One of the first people to work out the mathematical equations we use to describe the laws of the weather was born in Sligo in 1819. His name was George Gabriel Stokes and NWPs are based on his **equations**.

HISTORY

It was difficult to do accurate or quick forecasts based on equations. During the Second World War, work on developing **computers** really began in earnest, and by the 1950s, we had computers that were able to work out lots of different complicated equations. But they looked a lot different to the computers we're used to these days!

The first computer to be used to work out the weather was called the ENIAC. It had 18,000 vacuum tubes, 70,000 resistors, 10,000 capacitors and 6,000 switches. It was about the size of a really, really big house! It took too long – by the time it could work out the forecast, the weather had already been and gone. But it was a start, and we learned from it!

MET ÉIREANN'S MODELS

Now people use their phones to see what the weather is going to be like! Of course, the weather has to get into the phone first. We use our **NWPs** to do this. The main computer model used at Met Éireann is called HARMONIE. But we also use other models, including the ECMWF models and many others from around the world. We can also look at models from other parts of the world to get the full picture.

The models chop the area around the world up into boxes. When we're trying to predict what is happening in the boxes, we refer to them as grid points. We work out all the physics equations for each of the grid points and then use other equations to work out what happens in between. As computers get faster and better, we are able to increase the number of grid points we use for the model, which improves the accuracy. The data used by the model comes from weather measurements and satellite information. Once the data is put into the computer, a model of what the atmosphere looks like at the starting position is created. This starting position is then adjusted by the computer over time so we can see how the atmosphere is going to change.

ENSEMBLE FORECASTING

Each time the computer model runs, it checks with the latest observations to see how it is doing, and the model is then adjusted and run again. It is a good idea to run the same computer model over and over again – up to 50 times! – changing just one small thing in the starting position to account for anything we might have missed. This is called **ensemble forecasting**. When we look at what happens on the ensemble forecast, we can guess the most likely outcome by choosing the result that happens most often within the 50 trials.

After the equations have all been figured out, we use something called visualisation tools to draw pictures from the worked-out figures. The output from the computer models is used on Met Éireann's website, on phone apps and on the TV forecast.

BROADCASTING

Now that computer models have made forecasting the weather so much more accurate, how we share that information is more important than ever. This is called broadcasting.

MEDIA

We make weather forecasts on the television using videos, pictures and words. We also make weather forecasts for newspapers, which can use a collection of both pictures and words. On the radio, we only have words to explain the weather – so it's much easier on the TV! These days, people often get the weather forecast from the weather apps on their smartphones. We can also use social media to send messages to those who want to follow what the weather office has to say. We also broadcast our forecasts onto our website, as well as the latest weather, satellite pictures and radar images.

OBJECTIVITY

As weather forecasters, we must try our best to remove any subjective language from our forecasts and make the message as clear as it can be. Subjective means things like telling people it will be a 'nice' day. Well, some people like snow! So to them, a day is nice when it's snowing. Other people are terrified of thunderstorms, so that would be a bad day for them. But some people really love them, and even chase them around the country – we call these people storm chasers! So we should never say that it's going to be a nice day in the forecast. Instead, we must try to be **objective** and only describe the weather exactly as it is to the very best of our ability. We let the people decide for themselves if it's nice weather or not.

TV GRAPHICS

When making a television broadcast, we make up graphics that can be seen behind the forecaster. These graphics are made using the computer model. Once the graphics are made, they are sent to a computer. The forecaster will then go to the TV studio and stand in front of a blue screen. The TV camera is able to cut out all the blue in the recording, and where something blue shows up, the camera can put the computer graphics instead. We use a clicker to move through the charts, one at a time, and describe what the weather will do as each chart goes by.

Once the words have been recorded, the graphics are put together with the recording of the forecaster and both images are broadcast together over the TV. Did you know that if the forecaster wears something blue, they disappear? This is because the camera detects the colour blue and puts the graphic over where the forecaster is standing. So weather forecasters have to be careful not to wear blue. It's so much easier to broadcast on the radio – we can wear whatever colour we like there!

OPENING HOURS

Weather forecasting offices are open 24 hours a day and 365 days a year – even on Christmas Day! This is because most of the work that is done on a weather forecast has to happen before everyone is awake. People want to know what the weather is going to be when they wake up in the morning so that they can plan their day. They use this information to make simple decisions – should I wear a jumper or a rain coat? Should I carry an umbrella? Will I hang out the washing? There are also more important decisions – should we take the ferry? Should the farmer cut the silage? Should the pilot fly their plane? The weather forecaster works all night long, and the forecast has to be ready to broadcast on the radio or on the internet before 6 a.m. The rest of the day is then spent making sure the forecast is on track and any warnings that need to be issued are out to the public as quickly as possible.

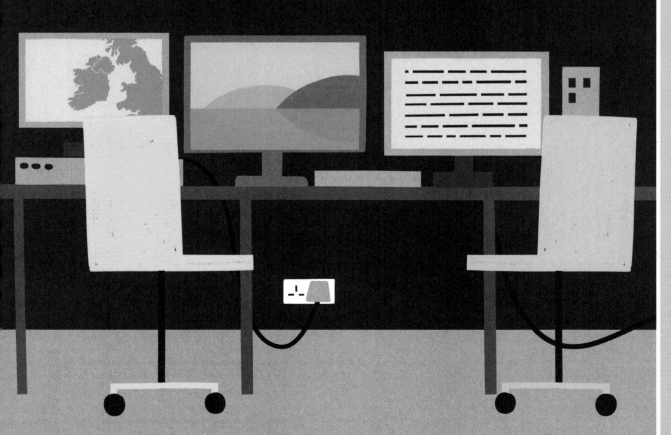

WEATHER CHARTS

Weather observers need to communicate the weather to others quickly and easily.
We use weather charts and special symbols to do this.

In the past, this was done by hand with information transmitted by radio signals. The weather was observed and a code was created to relay the information simply. This information was then decoded and plotted by meteorological officers and weather forecasters onto a chart so that a picture could be drawn at a glance.

In order for the weather picture to fit on a chart, symbols were created to tell the story. We still use these today. Here's an example of a typical chart.

The circle at the centre is used to tell us some information. The amount that is coloured in tells us how much cloud is in the sky.

The wind direction is shown by the direction of what we call the wind barb, and the number of lines crossing it tells us the speed. This is a northwesterly wind blowing at 23-31 km/hr.

On the left, we can see three dots – that tells me that there's heavy rain about. Above it is the air temperature, and the dew point is below.

We can also see from the curved line that the pressure is falling, but it will level off.

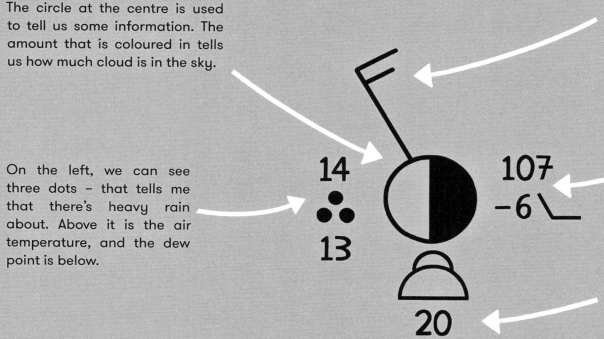

Underneath that, we draw a little picture of the type of cloud we see in the sky. This symbol shows that there are cumulus clouds nearby and the 20 underneath tells us what height they are at – 2,000 feet.

Weather, temperature, wind, pressure, pressure tendency, sky cover, dew point – that's a lot of information crammed into a little picture! All these symbols arranged around the centre circle make reading the chart very easy and clear for the forecaster.

PLOTTING SYMBOLS

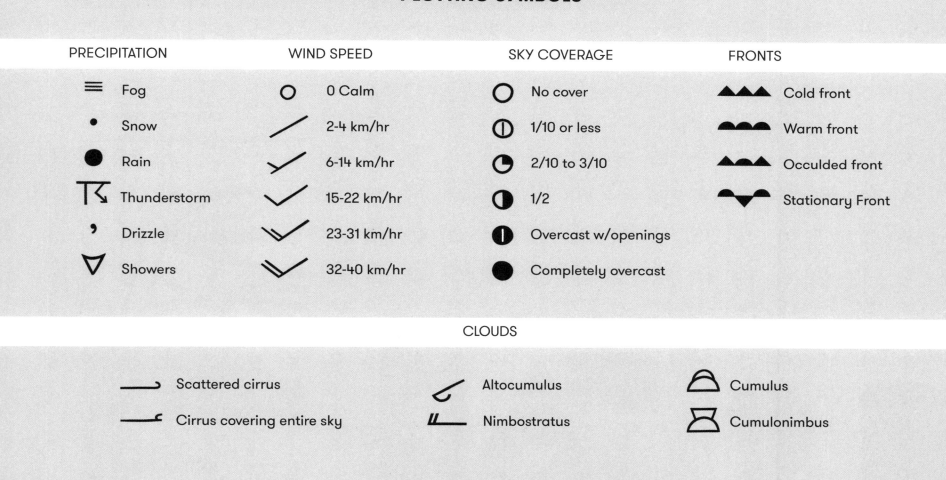

PRECIPITATION

≡	Fog
•	Snow
●	Rain
⊺⊼	Thunderstorm
'	Drizzle
▽	Showers

WIND SPEED

○	0 Calm
/	2-4 km/hr
∨	6-14 km/hr
∨	15-22 km/hr
∨	23-31 km/hr
∨	32-40 km/hr

SKY COVERAGE

○	No cover
◑	1/10 or less
◔	2/10 to 3/10
◐	1/2
◉	Overcast w/openings
●	Completely overcast

FRONTS

▲▲▲	Cold front
●●●	Warm front
▲▲●	Occulded front
▼▼	Stationary Front

CLOUDS

⌐	Scattered cirrus	⌐	Altocumulus	⌂	Cumulus
⌐	Cirrus covering entire sky	⌐	Nimbostratus	⌂	Cumulonimbus

What's the weather like where you are?
Can you write a forecast based on these symbols?

There is a weather symbol hidden on every page of this book – can you find them all?

NAMING STORMS

Naming weather systems is important for lots of reasons. It is easier for meteorologists to communicate and monitor a storm if everybody has agreed on the name. The public also loves to name storms, as they can prepare for them and track their progress.

HISTORY

In the past, storms were named for the year, location and sometimes for the things they destroyed. When a hurricane struck the Gulf of Mexico in 1842, it ripped the mast off a boat named Antje, so it was called Antje's Hurricane. In 1922, a storm destroyed the Knickerbocker Theatre in Washington DC, so it then became known as the Knickerbocker Storm!

In 1953, the United States began naming storms using only female names, and from 1978, both male and female names were used. These days, Atlantic hurricanes and tropical storms are named by the National Hurricane Centre under guidelines laid down by the World Meteorological Organisation. The list of names is drawn up in advance and alternates from male to female. When a storm causes large-scale devastation, the name associated with it is not used again.

Storms in other areas such as Australia, the Indian Ocean and the south Pacific also receive names from their local meteorological organisations. But until recently, winter storms that affected Ireland, the UK and Europe weren't given names. But things started to get confusing! Sometimes, when a big storm was on its way, the media would name it so that the story would grab the public's attention better. For example, the famous biologist Charles Darwin was born on 12 February 1809, so the storm that hit Ireland on 12 February 2014 was named Storm Darwin.

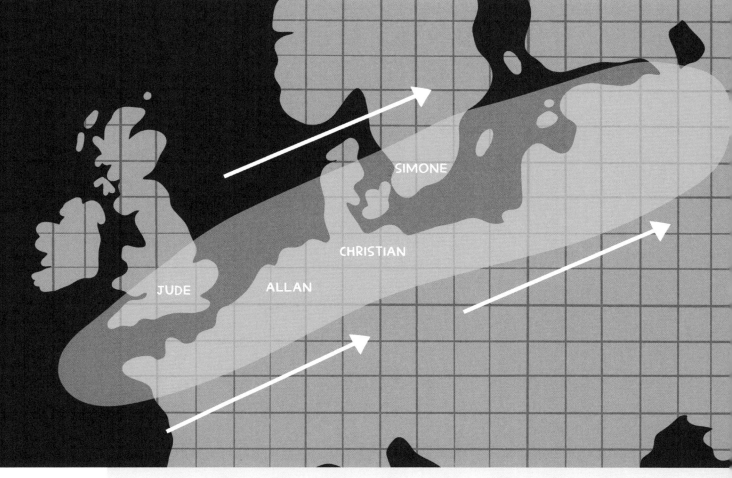

Another storm that hit in that same winter season reached its peak on St Jude's day and was called Storm Jude. But this storm affected more than one country, so as it moved through Europe it picked up other names too. It was called Christian in Germany, and in Sweden, it was referred to as Simone.

After all this confusion, the Met Office in the UK and Met Éireann got together and decided to start officially naming Atlantic winter storms together. The public was asked to contribute and the first officially named winter storm from the British and Irish list was Storm Abigail, which came on 10 November 2015. There are other partnerships across Europe – the Spanish, French and Portuguese name storms together, and in Scandinavia, the Swedes, Norwegians and Danes share their names.

You can pay to name a weather system at the Institute of Meteorology of the Free University of Berlin! If you 'Adopt a Vortex', your name will be attached to an area of low or high pressure on the European weather charts. An anticyclone will cost you more than a depression, though – they stay on the charts for longer!

WARNINGS!

One of the most important roles of a weather forecaster is to issue warnings to the public when dangerous weather conditions are approaching.

HISTORY

On 26 October 1859, a very bad storm passed through the Irish Sea between Britain and Ireland. It was named the Royal Charter Storm because it wrecked a great sailing ship called the Royal Charter. The ship was driven onto the shore by storm-force winds which increased to reach Beaufort force 12 on the scale – hurricane force! It was travelling from Melbourne to Liverpool and was carrying 450 people and a shipment of gold from Australia. The ship, cargo and all the passengers were lost at sea.

After the storm, Vice-Admiral Robert FitzRoy founded what later became the British Met Office. It was decided that some sort of warning system was needed to protect ships from storms. A series of beacons were established around the coasts of Ireland and Britain to be lit when gales were expected. These were the first weather warnings.

WARNING LEVELS

These days, warnings are issued on a scale. Yellow warnings mean that people should 'be aware'. The weather may not be widely dangerous, but to some people, the conditions may be dangerous due to where they are or what they are doing. The next level is more serious.

An orange warning means people must 'be prepared'. This is when the weather is expected to worsen significantly and people in affected areas should make sure that they are ready.

The red warning is for 'take action'. This is the highest level of warning and it means that there is a threat to the life and property of those in the affected areas and all precautions must be taken to stay safe.

TEMPERATURE WARNINGS

Met Éireann warns people when the weather is going to get very cold. This is because ice can form on the footpaths and roads, which can cause people to slip and cars to skid. County councils will put grit down to stop the ice from forming and to help protect the roads. We also issue cold weather warnings to homeless charities so that they can help people who are sleeping outdoors.

We also issue warnings if it is going to be very hot! If the temperature is high it can cause heat stress, which can be dangerous to people who are vulnerable. The sick or the frail may not be able to regulate their body temperature and they can become unwell or even die from the heat. Warm nights can be especially dangerous because our bodies don't get a break from the heat, so we also issue warnings for high night-time temperatures.

RAINFALL WARNINGS

We put out warnings if more than 25 mm of rain is expected in 24 hours, especially if the conditions are right for flooding to occur. Protection such as sandbags can be put out, and the engineers that look after our dams and rivers can be prepared for the water coming through the systems. County councils can make sure drains are cleared of leaves and rubbish and people can make sure they have extra time for any journey they need to make. We can't stop heavy rain from coming, but we can warn people!

WIND WARNINGS

Strong winds are probably the most common type of warnings that are issued. In Ireland, we get strong winds when storms or deep depressions come near or over the country. A yellow wind warning is issued when we expect winds to reach speeds of over 55 km/hr but less than 65 km/hr, with gusts up to 110 km/hr. Orange wind warnings are issued when winds are expected to reach up to 80 km/hr with gusts up to 130 km/hr. A red level warning – our highest warning – is issued when winds are going to be over 80 km/hr with gusts over 130 km/hr.

SNOW WARNINGS

Warnings in Ireland are also issued when we expect there to be accumulations of snow – even as little as 3 cm! Although it doesn't seem like a lot, snow can cause difficulty on the roads and footpaths, so people need to 'be aware'.

Other countries get heavier snow than Ireland, but because it happens so rarely here, we don't have the same systems in place to deal with it. Some people love snow and wait all year hoping for a snowfall warning, but it rarely happens!

OTHER WARNINGS

We issue warnings for fog too, but only at the orange level when we expect it to be dangerous. Similarly, we only issue thunderstorm warnings for the general public when we expect them to be very bad and dangerous.

There can be some factories and buildings that are vulnerable to lightning strikes and they need to know in advance if there is going to be a thunderstorm in their area – like people who make hospital equipment … or explosives!

EXTREME WEATHER

Hurricanes, tornadoes, monsoon rainfall and heatwaves happen seasonally across the globe but here in Ireland are a rare phenomenon that only happen once in a while.

Typically, extreme weather events happen in regions of the world where the temperature is at its highest. The heat from the sun's radiation makes for a lot of energy! For example, we see monsoon rainfall around the equatorial regions, and hurricanes develop over the warm Atlantic and Pacific Oceans. Destructive and violent tornadoes occur where the heat rising from the Gulf of Mexico meets the cold air rushing down from the great Rocky Mountains.

But extreme weather can happen anywhere. Despite our temperate climate – like Goldilocks' porridge, not too hot and not too cold – Ireland has had its fair share.

THE NIGHT OF THE BIG WIND

It had snowed the day before the storm, so it was colder than usual. On the morning of 6 January 1839, an Atlantic depression deepened, driving a warm front over the country. This rapidly melted the snow and it became warm for a time. There was an eerie calm – the air was so still that people could hear voices from houses a mile away.

The melting of the snow tells us that there was a big temperature difference between air masses somewhere in the area. When there is a big difference in air masses, there's a lot of moving air. Sure enough, the wind started to increase as a cold front came in behind the warm front. People could sense that something large was approaching, and many could hear a rumble coming from the western horizon.

Stormy weather arrived on the coast of Mayo at around lunchtime, and the wind strengthened all day. By midnight, the winds had reached hurricane force. Waves broke over the top of the Cliffs of Moher, huge rocks were thrown onto the Aran Islands, and all the water was blown out of the canal at Tuam. Thatch roofs were blown away, windows were broken and trees were ripped up.

Long before hurricanes had names, and even before there was a Met Office in the UK, some storms were so strong they went down in Irish folklore. The Night of the Big Wind (Oíche na Gaoithe Móire in Irish) was a huge storm that hit Ireland on 6 January 1839, and stories have been told about it ever since.

People were terrified and tried to find shelter anywhere they could. There was no electricity at the time, and the wind would have blown out all the candles and lanterns, so it all happened in the dark! How scared the people must have been, not knowing what was happening to the world around them.

Most of the damage was seen in Connacht, Ulster and north Leinster, as the storm swept across the northern half of the country before moving away to northern Europe. It is reported that about a quarter of all the houses in Dublin suffered some damage, but inland and across the country, where dwellings were less solid, there was total devastation. Houses were destroyed, farms flattened and roads were blocked all around the country.

Extensive flooding was reported as storm surges swept seawater miles inland, which caused a knock-on effect on farmland and crops even after the storm. So many trees fell that there was no place for the birds to nest, so there were no birds singing that spring. Because so many people were not registered at birth, there is no accurate account of how many people lost their lives on the Night of the Big Wind, but it is estimated at hundreds.

WHERE WERE YOU?

Where were you on the Night of the Big Wind? When the old age pension was first established in Ireland in 1908, this question was used to assess who was old enough to receive it. A lot of poor Catholics weren't registered at birth, so they didn't really have a way to prove that they were old enough to have a pension!

The storm happened on 6 January, the religious feast day of the Epiphany. The Catholics of Ireland were a very superstitious lot and many felt that the storm was signalling the end of the world. Others said that this was the night that the English fairies invaded Ireland and defeated the native Sidhe!

HURRICANE DEBBIE

Hurricane Debbie came in 1961 and is one of the most devastating storms to hit Ireland in living memory. The storm began to develop in Central Africa in late August. By the time it reached the Cape Verde Islands on 6 September, it had reached hurricane status, and the strong winds caused a plane to crash, taking the lives of 60 people.

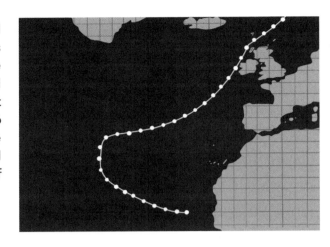

The storm then started to move westwards towards the Caribbean, which is the usual course for storms of this nature. However, about half-way across the Atlantic, it took a right turn and started to head north eastwards, heading towards Ireland. We didn't have the satellites then that we do now, so we didn't know it was coming. It was only by chance that an aeroplane was flying over the area and spotted the hurricane on its way to Ireland!

The storm reached Ireland on 16 September 1961, and the first winds arrived on the west coast in the morning. Right at the very top of the country, at a place called Malin Head in Donegal, the anemometer recorded a gust of 181 km/hr. When you consider that the fastest you can drive your car on the road in Ireland is 120 km/hr – that's a strong wind!

Tens of thousands of trees were knocked down and buildings were badly damaged. The Galway Market was closed because forecasters predicted destruction – and they were right! The city looked like a war zone after the hurricane went through, with debris from damaged buildings everywhere. In Cork airport, the control tower windows were blown out.

The storm caused the deaths of 12 people in the Republic of Ireland and another 6 in Northern Ireland. It is estimated that the damage to buildings, trees, farms and crops cost between 40 and 50 million euro. The storm kept going through the north of the UK, over Norway and into Russia, and it took the lives of 78 people in total.

HURRICANE CHARLEY

In August 1986, the tail end of Hurricane Charley came to Ireland. A hurricane needs warm sea-surface temperatures to keep it going, so when it moves into cooler waters, it starts to transition into what we call an ex-hurricane or an ex-tropical storm. Hurricanes are like presidents – even when they're not hurricanes anymore, they get to keep the title, they just lose some of their power!

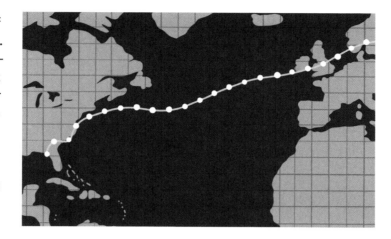

Hurricane Charley started its life as a storm off the coast of South Carolina in the United States on 15 August 1986. The hurricane brought flooding, fallen trees and car accidents, and it even caused a small aeroplane to crash. Hurricane Charley was already devastating when it finally made the transition to ex-tropical storm over the Atlantic. But it wasn't finished yet!

On 25 August, ex-hurricane Charley arrived on the south coast of Ireland. While the storm brought strong winds, it is the heavy rain that Charley is remembered for. The storm broke all previous rainfall records for Ireland – over 200 mm fell in a 24-hour period. Met Éireann issues rainfall warnings if more than 25 mm of rain is expected in 24 hours, and this was almost ten times that amount! Rivers like the Dodder in Dublin and the Dargle in Bray burst their banks, and there was widespread flooding. In Bray, homes were flooded to a depth of 1.5 metres – that's almost as tall as me! In Dublin, flooding was reported up to a height of 2.4 metres.

When rescuers arrived to help people evacuate, they needed to be taken to safety by boat. In the English Channel, a boat carrying 31 passengers was flooded when it was hit by an 8 metre wave, and the people had to be rescued by helicopter. In total, 13 people lost their lives because of Charley.

Because the storm flooded so many fields, lots of crops were lost that summer. After the storm, the government of the time set out £6.5 million to repair the roads and buildings damaged by ex-hurricane Charley.

Debbie and Charley might sound like your friendly neighbours – but they weren't very friendly to Ireland!

FLOODS

Whether or not **flooding** occurs after heavy rain depends on the conditions and time of year. If a lot of rain falls in August after a dry summer, there is plenty of space in the earth and rivers for the water to flow away. But if the same amount of rain were to fall in November, the rivers and ground may be already saturated due to autumn rain, and the drains may be full of fallen leaves from the trees.

If the rain is very heavy and comes all at once, it can pool in puddles causing what we refer to as local or 'spot' flooding. If it falls slowly over the course of a whole day, there is time for the rain to wash away and it causes no problems at all – except lots of wet feet!

FLOODING IN IRELAND

Flooding happens every couple of years in Ireland, but some years are worse than others. Scientists at Queen's University Belfast recently uncovered evidence that about 4,000 years ago in Ireland, it rained for 20 years straight! Around this time, a volcanic eruption caused the earth to cool, and there was flooding all over the world. Something similar happened again about 2,000 years ago, only this time it rained for ten years straight. We haven't had anything like that in recent history – perhaps we're overdue. Maybe we should learn how to build an ark!

However, we have had some bad flooding in Ireland over the last 20 years. Most rivers have what are called **flood plains** along the banks of the river that take up extra rainwater. But when Dublin city started to grow during the building boom of the late 1990s, the demand for houses pushed estates onto the flood plains of the Tolka River. When heavy rain came in 2002, the river couldn't use its flood plains to get rid of the extra water. It had to go somewhere, and this time it ended up coming through the front doors of unsuspecting Dubliners – 200 people had to be evacuated!

WELCOME TO
KINVARA

In Ireland, when it rains very heavily we say 'it's lashing out!' But how much rain is this? Well, on a rainy day in Dublin, we might get about 7 mm of rain. If you ever decide to visit India during the monsoon season, you might see what lashing really means – 200 mm of rain could fall in a day! Fortunately, we don't get monsoon rains, but we do frequently have flooding.

Coastal flooding can happen when there is a high tide, especially if the wind is driving the waves towards the shore at the same time. A high tide will also stop the rivers from being able to empty into the sea.

Flooding can also occur after snow. As the snow melts, it will filter down to the rivers, which can get overwhelmed if the snow melts too quickly. After periods of snowfall, the government will call on weather forecasters to predict how long it will take to melt, so that they can judge if rivers are likely to burst their banks.

Another wet autumn came in 2008, when, for the first time in the history of Croke Park, the sky was so dark during the day that floodlights had to be used for the All-Ireland football quarter-final. In November 2009, Valentia Observatory recorded the wettest month ever, and their records go back to 1866. And in 2014, in Kinvara, the quays flooded so badly that some fishing boats were floating in the streets. Maybe we do need that ark!

THE ASH CLOUD

Sometimes, weather forecasters are asked for advice in areas that
are not necessarily meteorological, for example, when a volcano
on the island of Iceland erupted in 2010.

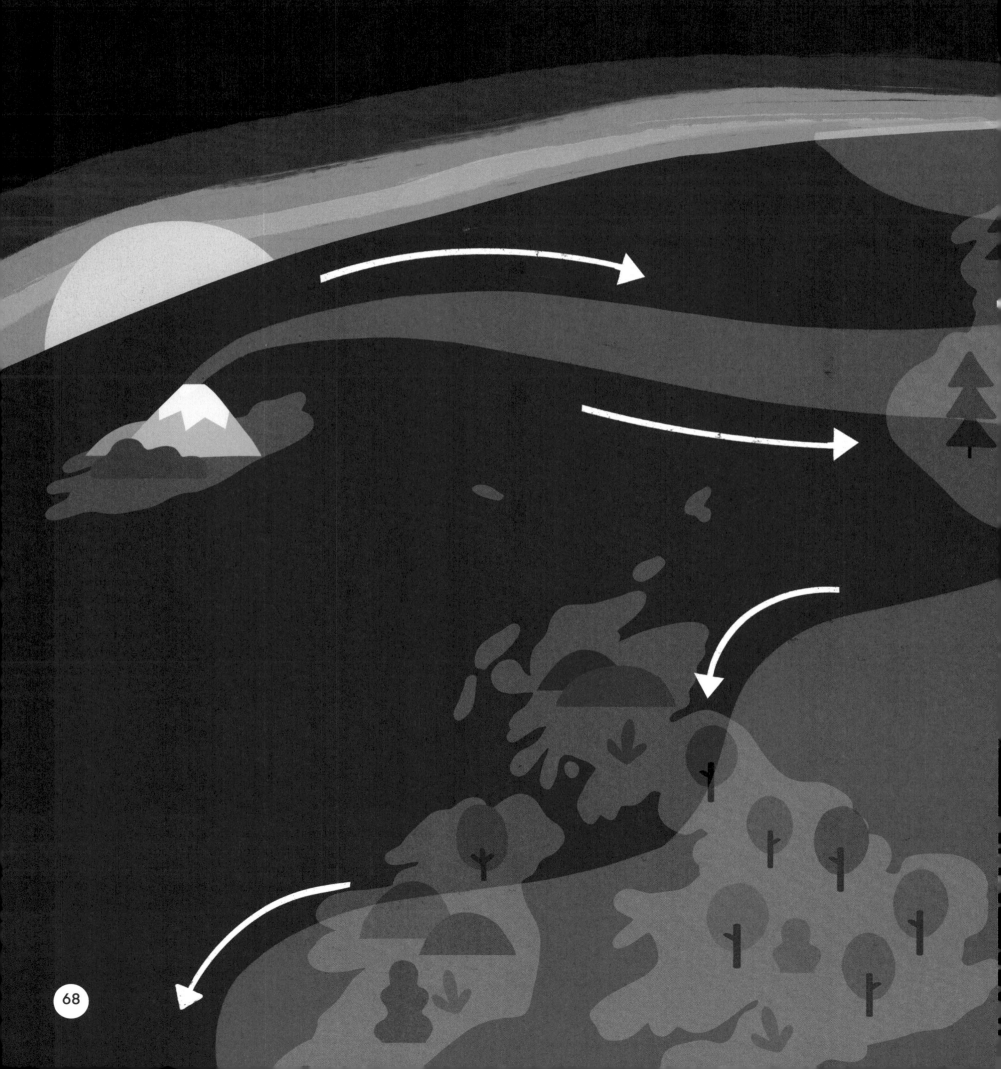

The winter of 2009–2010 was already disruptive. Devastating floods had occurred in November 2009, and snow and ice came along later. The forecasters at Met Éireann had already been challenged by a tough year when a volcano called Eyjafjallajökull erupted, sending clouds of ash into the atmosphere and causing major problems for air traffic all over Europe. (It's a tongue twister – try saying 'ay-yah-fyah-plah-yer-kuh-duhl' ten times!)

When a volcano erupts, the force of the explosion pushes huge amounts of volcanic ash up into the troposphere and beyond. The ash cloud is made up of burning rocks, magma, glass and gases. Some are large pieces that fall to earth quickly, and others are microscopic – we can barely see them with the naked eye. When the ash starts to fall back to earth, the smaller pieces get carried along with the wind. The wind direction and strength will determine how far the ash is carried.

When Eyjafjallajökull erupted in 2010, the position of the areas of high and low pressure meant that a northwesterly wind put Europe and Ireland directly in the firing line! This was when Met Éireann's forecasters came in. They looked at the developing weather patterns to determine where the ash would go, using special computer programs called **dispersal algorithms**. These are used when we want to see what would happen to particles travelling in the air high above the ground. The computers work out how the particles will be mixed up, spread out and carried on the breeze. We use dispersal algorithms to track where anything in the air might go, such as pollution, disease, chemicals – and ash clouds.

Burning ash from a volcano is extremely dangerous to aeroplanes. The ash is sucked into the engines and could cause the plane to crash. Because of this danger, the airspace above Europe was closed to all flights between 15 and 23 April. This was the longest shutdown of the airspace since the Second World War! Millions of people were stranded across Europe and the whole world as flights were cancelled. The airspace was opened again once the wind direction changed, but later closed for shorter spaces of time over the next month as the eruptions continued.

The tiny particles of ash in the air caused the light to bend and made the sky look red, pink or orange in many places. People reported seeing beautiful sunsets during the eruptions. But they also reported smelling rotten eggs, caused by the sulphur in the air. Yuck!

HURRICANE OPHELIA

In 2017, this powerful storm hit Ireland – and this time, the forecasters at Met Éireann were ready.

Ophelia, like Debbie, started out as a tropical storm that formed off the coast of Africa from a dying cold front on 6 October 2017. The oceans in the area were particularly warm for the time of year, giving Ophelia extra energy. It became a hurricane on 11 October and intensified into a major hurricane when it moved past the Azores, bringing high winds and heavy rainfall to the area. The storm was still a hurricane as it started to move closer to Ireland, but it made the transition to an ex-tropical storm early in the morning of 16 October. Instead of weakening, it gathered strength as it approached the south coast of Ireland, interacting with strong southerly winds.

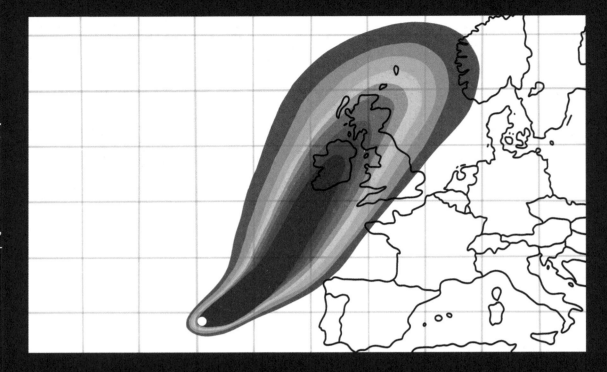

MET ÉIREANN

Back at Met Éireann's headquarters, we had been in contact with the National Hurricane Centre in the United States, along with the British Met Office and the National Weather Service in Washington DC. The computer models behaved fantastically, giving us very strong confidence in the expected path of the storm across the country.

As the storm began to move northwards, a lot of information began to flood the media, as online enthusiasts got their hands on charts from the National Hurricane Centre. The public was starting to become aware – but also confused. It was our task at Met Éireann to deliver an authoritative voice on what could be expected as the storm hit.

As one of the forecasters on duty, it was my job to start to prepare the people of Ireland for the impact of a major hurricane. New schedules would need to be adopted instead of the normal forecasts – there would be no farming forecast or world weather, only Hurricane Ophelia! Ophelia was forecast to arrive early on 16 October, with the eye of the storm passing along the west coast. The heaviest of the rain comes to the north and west of the eye of the storm, so thankfully much of it was expected to go out to sea. But not all of it, so heavy rain was still forecast for the north and west of the country. Elsewhere, there would be strong southeast winds at first, then strong southwest winds as the storm drove northwards.

The decision was made to issue warnings earlier than usual in the run-up to the event. The storm would hit early on Monday morning, and people would be leaving work on Friday so they needed enough time to prepare. The red-level warnings were first issued to the coastal counties of the west, but during Sunday the decision was made to extend the warnings to all areas of Ireland. We were dealing with a hurricane, and we needed to keep the people of Ireland safe. The red-level warning had the desired effect – schools and universities closed, buses were cancelled and many businesses decided to close. Most importantly, people made the decision to stay home.

RTÉ OPHELIA

ALERT

RED LEVEL WARNING

PLEASE STAY INDOORS

THE STORM

Our equipment was ready to monitor the conditions as the storm approached. When Ophelia reached Fastnet Rock off the coast of Cork, the strongest wind ever reported in Ireland was recorded, with winds gusting up to 191 km/hr. The weather observation buoy at M5 off the south coast (the M stands for marine) was knocked out of operation. Our climate station at Sherkin Island in Cork was also damaged early in the day.

A strong, dry southeast wind swept through the whole country first as the winds revolved anticlockwise around the eye of the storm. People in Kerry reported clear blue skies and dead calm conditions as the eye of the storm passed overhead. Later on, the winds would veer to the southwest as the storm moved up the coast.

Heavy rain pushed over the north and west as the storm moved northwards later in the day with Donegal reporting areas of spot flooding in the heavy downpours. Strong winds battered most of the country, but thankfully, most people stayed safe indoors.

Trees and power lines were knocked down all over the country, causing more than 360,000 homes to lose power. A stadium roof in Cork collapsed and a school roof was recorded flying through the air. Tragically, three people lost their lives during the course of the day.

Gardeners across the country noticed that many plants had been destroyed by the strong, dry wind that swept through the country. The leaves were burnt dry!

In the UK and across Europe, an eerie orange sky was described and the air seemed to be filled with dust. The winds of Hurricane Ophelia had swept sands all the way from the Sahara Desert over the whole of Europe.

THE BEAST FROM THE EAST

Why don't we get regular snowfalls every year like they do in most other countries that are at a similar latitude? The reason for this is that the Gulf Stream and the jet stream tend to keep the weather here in Ireland from getting too cold.

THE BEAST

In February 2018, meteorologists started to notice a disturbance in the stratosphere that signalled that the jet stream was set to reverse in direction, allowing a huge area of high pressure to build over Scandinavia. Because winds go clockwise around areas of high pressure, the position of the anticyclone was set to put easterly winds over Europe heading straight for Ireland. Areas of high pressure are big and slow, so they are much more predictable than depressions. Meteorologists could forecast this weather development well in advance – it was looking like snow for Ireland!

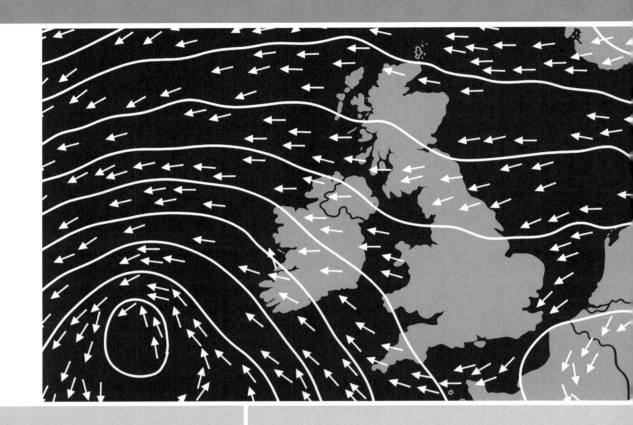

Snow showers started to affect the UK, and soon, people in Ireland were being warned to expect the same. The first weather warning went out the weekend before the snow was expected, giving people plenty of advance notice. People listened to the warnings from Met Éireann and stocked up on essentials at the supermarket, making sure they had plenty of fuel and preparing themselves for a snow day.

On 28 February, some light and scattered snow showers started to arrive. But as temperatures fell that night, they became much more frequent. Warmed by the Irish Sea, the air rose and condensed into great shower clouds. When they met the very cold air over the land, they fell as snow all along the east coast of Ireland.

The **jet stream** is a band of strong winds that forms when the cold air from the north pole and the warm air from the tropics meet. It lies very close to where we are – at 53 degrees north of the equator. Usually, the winds come from the west or south, from the warm Atlantic Ocean. Sometimes, the jet stream can become disturbed, and the winds get mixed up and start coming from the east. When this happens in the winter, it gets very cold over Europe.

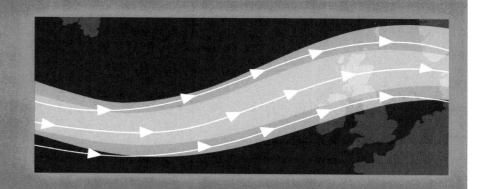

Temperatures started to fall day by day until they were 5 degrees below what they usually are in February or early March in Ireland. Our neighbours across Europe were starting to feel the effects. People in the Netherlands called it the Siberian Bear and looked forward to being able to skate on the canals. In the UK and Ireland, the weather became known as the Beast from the East.

The next day, schools were closed in many parts of the country – mostly in Leinster and Munster – as roads became dangerous due to lying snow. That night, the showers became steadily worse and more widespread. Pushed by strong winds, they were carried right across the country as far as Clare and the Inishowen Peninsula. In Mayo, temperatures fell to minus 6 degrees in the air, with steady winds of 40 km/hr over land – that's cold!

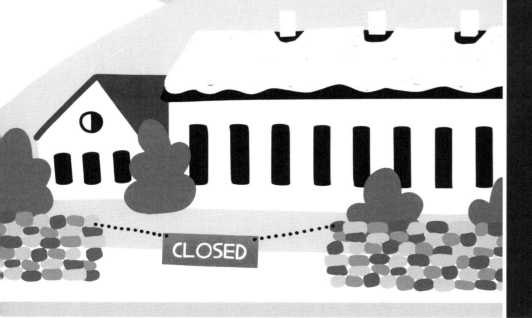

STORM EMMA

The worst was yet to come. An area of low pressure was developing to the south, and the Portuguese Met Office had named it Storm Emma. Storm Emma was developing in the warmer waters to the south of Ireland and west of Portugal, and it was moving north towards Ireland and gathering strength. It would bring storm-force winds, but more than that, it would bring a lot of moisture. Because Storm Emma was travelling over the warmer oceans, it was picking up lots of water vapour. When that air met the cold conditions from the east, all of it would fall as snow.

On 1 March 2018, Storm Emma arrived in the south of Ireland, and gale-force winds drove the snow into drifts right across Leinster and Munster. People reported opening bedroom windows in Naas and seeing the snow right up to the windowsill. Cars were buried all over Wicklow and Wexford and people were snowed in for days on end! Thankfully, people had listened to the warnings from Met Éireann and were prepared.

PEOPLE AND WEATHER

People are fascinated by the weather and always have been. Scientists all through history have worked on the problem of how to predict something that appears to be so unpredictable, using maths and physics to create hugely complicated equations. Each century, scientists stand on the shoulders of those that went before them and build on their work – and that's still going on today.

But it's not just scientists who are interested in the weather. It affects how we do our jobs, how we travel and how we live our lives. Everywhere I go in the world, almost every conversation starts with a comment about the weather. It's not just Irish people – although some suspect we are more obsessed with it than others!

ROBERT BOYLE

Robert Boyle was born in Waterford in 1627. While still a student, he travelled to Florence in Italy and studied the work of a stargazer called Galileo Galilei, beginning his love of science. On his return, Robert began down the path that would lead to him being thought of as the first modern chemist. He is best known for Boyle's Law, which says that if you have a fixed amount of gas, kept at a fixed temperature, the pressure and volume are proportional. As we have learned, gas and pressure are key parts of meteorology, so his work has helped future scientists understand the weather better.

FRANCIS BEAUFORT

Sir Francis Beaufort was born in Navan in 1774. Francis' father was a clergyman and a map maker, and he published a new map of Ireland when Francis was a young man. Francis left school early and went to sea, but at age 15 he was shipwrecked due to a faulty map! This led to his lifelong dedication to marine charts and weather reporting. Despite leaving school early, Francis was highly educated and well respected, and he became an officer in the Royal Navy. He developed the Beaufort Scale for describing wind speed, and it is still widely used today.

GEORGE GABRIEL STOKES

George Gabriel Stokes was born in 1819 in Sligo. Stokes studied mathematics in Cambridge and afterwards worked as a professor there. In December 1879, a winter storm destroyed a bridge in England while a train was crossing, and everyone on the train died. Stokes was appointed to study the effects of strong winds on high structures. He earns a spot in the meteorology hall of fame not just for his work, but also for his role in the Royal Society. His name lives on in the Campbell Stokes recorder, which was used to record sunshine hours until very recently. This was invented by John Francis Campbell and later refined by George Gabriel Stokes.

From as far back as the 17th century, Irish men and women have studied and modelled the laws of mathematics, physics and chemistry, helping to improve the science of meteorology.

WOMEN AND WEATHER

Amongst all these Irish men we could very well ask: Where are the women? Until recently, women were not encouraged to pursue education. Even when they made important mathematical and physical discoveries, their male colleagues often took their work for themselves. For this reason, many female scientists are often forgotten or ignored. But there were plenty of Irish women who were important contributors to the field of meteorology.

ANNIE MAUNDER

Annie Maunder was born in Tyrone in 1868. She studied magnetic changes on the surface of the sun and how they affected the climate of earth. She was elected a Fellow of the Royal Astronomical Society, and a crater on the moon is named after her and her husband!

KATHLEEN ANTONELLI

Kathleen Antonelli was born during the War of Independence in a Gaeltacht area of Donegal and later moved to the United States. Kathleen was a brilliant mathematician and physicist who worked on the ENIAC project, which was one of the first computers used to predict the weather.

ANNA DOBERCK

Anna Doberck was born in Copenhagen but came to Sligo in 1874 with her brother when he was assigned to work at the observatory at Markree Castle. Her observations were some of the most meteorologically accurate at that time, and she sent reports twice daily to the new British Met Office. She moved to Hong Kong to continue her work, where she earned the nickname Typhoon Annie – not because she caused them, but because she was so good at recording their activity!

LORD KELVIN

William Thomson, also known as Lord Kelvin, was born in Belfast in 1824. William was a keen mariner and he worked on the transatlantic telegraph project – the link that joined the 'New World' of the United States with Europe. This made almost instant communication between the continents possible for the first time. William also played a key role in discovering the value of absolute zero. It was known there was a limit to how low temperature could go, but he discovered that the lowest value was -273.15 degrees Celsius. The temperature scale of Kelvin was named in his honour.

RAY BATES

More recently, an Irishman helped to develop the Numerical Weather Prediction computer models that we use today. Professor Ray Bates worked at Met Éireann between 1963 and 1987. He went from there to NASA, where he stayed until 1995, and then back home to Europe to be a professor in Copenhagen. He is now an Adjunct Professor in Meteorology. Professor Bates received an award from NASA for developing new weather prediction models. We are very proud of the work of Professor Bates!

These are just some examples of Irish women who have made contributions to the field of meteorology. These days, there is no stopping the advancement of women, and the new Head of Forecasting at Met Éireann is female.

FOLKLORE

Folklore is the way people passed on useful information before there were written records. Knowing the weather was important to people's day-to-day lives, and they sometimes made up sayings to make things easier to remember.

'Red sky at night, sailor's delight. Red sky in the morning, sailors take warning.'

This is probably one of the oldest pieces of weather lore, with references to it even appearing in the Bible. The idea behind this little rhyme is that sailors – or shepherds – need to know what weather is coming their way. This rhyme was one of the earliest weather forecasts, and it mostly works!

In many places, weather systems come in from the west. As the sun also sets in the west, the sky at night gives us a hint of what will come the next day. When pressure is high, dust and sand get trapped in the atmosphere. This causes the blue lightwaves to scatter away, leaving only red light visible from the ground.

If we see this red light in the sky at sunset, that means high pressure is somewhere out there, bringing fine weather our way the next day – a delight! The opposite, a red sky in the morning, means that the high pressure is moving away to the east, allowing low pressure – which brings wet and windy weather – to come our way.

'Halo rings moon or sun, rain's approaching on the run.'

This refers to a halo that can sometimes appear to shine around the sun or the moon. It's caused by very high cirrus clouds that are almost invisible and mostly made from ice crystals. This type of cloud comes early in the progress of a warm front, which usually brings at least some rain and more unsettled weather later. This is also a useful forecast – sometimes! It's nice to know that a front is approaching, but the rhyme doesn't tell us when the rain will come or if it is sure to head our way. But it's good to get a heads-up if there's nothing else available!

'Rain before seven, fine by eleven.'

This one depends on the fact that in Ireland weather fronts tend to move through fairly quickly. If we have rain in the morning, it's usually gone within a few hours. It doesn't necessarily work as well on the continent of Europe where rain fronts take a little longer to move through.

'It's too cold to snow.'

Not true. It can snow at any temperature once there's enough moisture in the air. However, it is also true to say that warmer air can hold more moisture, allowing more precipitation. The colder the air, the lower its ability to hold moisture, and if the air doesn't have any moisture in it, well, it can't snow.

'Pine cones open when good weather is on the way.'

Pine cones open and close depending on how dry the air is. So when humidity levels increase ahead of rain, they are more likely to close, and when it becomes drier, they open up. Not a huge amount of advance warning with this one, but still, better than nothing in a push!

'Lots of berries mean a harsh winter.'

The idea here is that extra berries appear in autumn to provide for the birds in the cold winter ahead. The problem with this theory is that the berries are there because of the spring weather, not the weather that's due to come in the following three months. It's nice for the birds, but nothing to do with the weather!

'Cows lie down when it's about to rain!'

Do they? There's no scientific or meteorological explanation for this. People suggest that cows might be more sensitive to changes in pressure and wind, or that they can smell rain. I find it highly unlikely, but I would never argue with a cow!

WEATHER AND NOSTALGIA

Have you heard of Pollyanna? Pollyanna is a fictional character who first appeared in a book written in 1913 by Eleanor H. Porter. The character of Pollyanna is a happy and cheerful girl who spreads her optimistic outlook through her small town in the United States. The book was so successful that the name of the character, Pollyanna, has become associated with someone with a sunny personality.

In 1978, scientists showed that the brain is more likely to store memories that are pleasing. This has become known as the Pollyanna principle – people tend to remember happy, cheerful memories more clearly than sad or unhappy ones. Not only do people recall positive experiences more clearly, but they are more likely to store the information in the first place.

And that goes for the weather too – people remember the heat waves, the snow days, the ice creams at the beach, the snowmen on the lawn. They don't remember the dreary days stuck inside waiting for the rain to stop!

People often tell meteorologists that the weather was hotter when they were younger, or that it snowed more twenty years ago. They tend to get a bit grumpy when we explain they are wrong!

Another reason for weather nostalgia is old photographs. In the past, if you had a camera at all, taking a photo could be expensive and difficult to do. Photos were only taken sparingly on special occasions – on a summer holiday, a trip to the beach or on a particularly snowy day. We didn't take photos of the rain so much! Because of this, we appear to have lots of photos of sun and snow in the past. Every summer was sunny and every winter was snowy. Which, of course, is not true at all. Nowadays, we can take as many photos as we like with our phones!

The data suggests that the long-term averages of weather for Ireland have not changed dramatically over time. This is why it is so important to keep scientific records – unfortunately, people can get things wrong.

There have been sunny summers – ask someone old enough if they remember 1976! Although I was very young, I remember the sun vividly. I remember the sunburn too! And there were snowy winters – ask the same person where they were when it snowed in 1982. And there were storms, like Hurricane Debbie in 1961. And floods, like Hurricane Charley in 1986.

The records of weather go even further back. We know that there was an exceptionally fine and sunny summer in 1798 and very heavy snow in 1917. The warmest summer on record is 1995 when an anticyclone set in over the country for months on end and the sun shone every day. But then there was another really warm summer 11 years later in 2006. In 2018, Ireland saw another heatwave in June and July. How will children remember the summer of 2018 in twenty years?

JOBS AND THE WEATHER

PILOTS

Telling the public that it will be a cloudy day is enough information for most people. But it's not enough for someone flying an aeroplane! This is where aviation forecasters come in. The main forecast for pilots is called a terminal area forecast, which gives the weather for the area immediately around the runway of the airport. Take-off and landing are the most dangerous parts of the whole flight, so this is important for a pilot.

Different clouds have different properties, and many can be dangerous to fly through. Large cumulonimbus clouds have downdraughts that can knock even a jumbo jet from the sky. If there's going to be cumulonimbus clouds in the way of an aeroplane when they are coming in to land, they need to know about it!

The aviation forecaster will also forecast a description of the atmosphere for the area of the sky they are responsible for. They will also show on the forecast where there are fronts, turbulence and ice. It's dangerous for planes to fly through ice, because it can disrupt the air flow around the wings, making it difficult to fly. Aviation forecasters also have to warn about dangerous weather situations so that pilots can make plans to avoid bad weather and divert to different airports.

FARMERS

Farmers rely on weather forecasts to plan their work on a daily basis. The farming community needs to have information on rainfall, temperature and wind to prepare food for their animals, water for their plants and look after their farm's machinery. The soil temperature is very important to the farming community as grass will only grow when the temperature of the soil is above five degrees. In Ireland, soil temperatures tend to remain high enough all year round for grass growth and only slow down during the very depths of winter.

Rainfall totals are very important too. Apart from needing rain to water their fields, farmers need sunny days in order for grass to dry out to make hay. The fields also need to be dry enough for cattle and sheep to walk on. Waterlogged fields will turn to mud if animals are let out too soon in the spring. During lambing season, farmers need to know the temperature, as lambs are particularly susceptible to cold weather. Windy weather also needs to be forecast to make sure farmers are able to spray their crops as they cannot spray when winds are high.

Potato blight is also something farmers watch for. During the 19th century, a million Irish people died during a period known as the Great Famine. Many poor Irish people depended solely on the potato for their food, and a sickness called potato blight caused many crops to die. Met Éireann monitors the weather conditions so we can warn farmers if potato blight is likely to happen.

Nearly everybody's job is affected by the weather in some way. People need to dress in the right clothes every morning, travel to school or work, and have something to talk about at lunchtime! But some jobs rely on weather forecasts more than others.

MARINERS

The safety of anyone who goes out on the sea depends on the weather. Mariners must know the current meteorological situation so that they can judge what the weather might be like during their planned journey.

A sea area forecast will consist of wind strength, direction and overall visibility, along with a report on sea conditions and high waves. Mariners are particularly susceptible to high winds, so gale warnings are issued by meteorological services around the world. As well as a gale warning, a small craft warning is also issued for slightly lower winds. This is for smaller and more vulnerable boats.

The history of the meteorological service owes its development to mariners' dedication to safety at sea. To this day, weather reports are recorded by ships to report back on conditions to other people waiting on land.

ROAD USERS

Other customers of the meteorological services are those that look after the roads. County councils around the country are responsible for keeping our roads safe to drive on. Weather forecasters send the county councils a special forecast every day in winter to let them know if the roads are at risk of becoming icy.

When it is particularly cold, they send out the gritters! The gritting trucks spread a special kind of salt on the road that helps clear the ice and stop it from coming back. Be careful not to be on the footpath when one of these trucks goes by – you don't want to get a mouthful of sand and salt!

WEATHER SAFETY

Whether the weather is cold, whether the weather is hot, we have to put up with the weather, whether we like it or not! We can't avoid the outdoors, but we can be prepared. Met Éireann's job is to make sure that the public is aware of any dangerous conditions, but you can keep yourself safe by taking some simple steps.

Some people say that there's no such thing as bad weather – just bad clothing! Stay warm when it's cold by wearing lots of thin layers that you can remove if necessary. Keep your extremities warm – that means your fingers, toes, head and nose!

When it's hot, you should wear a hat and make sure your clothing is nice and light. To avoid sunburn, always wear sun block and reapply regularly through the day, especially if you have been swimming. If you're out in hot weather, drink lots of water to avoid becoming dehydrated.

If there's been heavy rain, stay away from rivers as the water will be higher and faster than usual. Never, ever drive into a flooded road. It could be deeper than it looks! You never know if you will get caught up in a current and washed away. Instead, find high ground and stay away from the water.

If you find yourself out in a thunderstorm, don't shelter under a tree! Trees and metal objects act like lightning rods, so it's safer to keep away. Keep low to the ground and get inside a building as fast as you can.

When out in the snow or on the ice, you should walk like a penguin to stop yourself from slipping. Lean forward, put your arms out and take small, penguin-sized steps. You might look silly, but not as silly as if you fell over!

If you are climbing or walking in the mountains or forest, make sure you have the necessary equipment. Always let someone know where you're going and when you expect to be back. The weather can change quickly in the mountains, and don't forget that the temperature drops the higher up you go!

If you're at the beach, watch out for currents, swim close to the shore and always wear a life jacket out on the water. Stay away from the coasts during high seas and stormy weather, because people can be very easily swept out to sea. Don't go for walks along the shore, even just to look at the waves!

During very bad weather, check in on elderly or sick neighbours to make sure they have everything they need. They might need food, fuel or just a chat, and a helping hand can make all the difference.

Remember the emergency numbers – call 112 or 999 if you see someone in trouble. You'll need to know your location and the type of help you need. Don't assume someone else will call!

GOODBYE

I HOPE YOU'VE ENJOYED reading my book as much as I've enjoyed creating it for you. At the start, I said that if you didn't like science, then perhaps you just hadn't been properly introduced. I hope I've done a good job of introducing you to one of my very favourite science subjects – meteorology.

This book is not children's science or a watered-down version of the truth. All of the facts are scientifically accurate to the best of my knowledge and experience. The answers in the book are to questions that adults ask me all the time. If you're an adult reading this and you have found it interesting, don't be surprised!

I hope you try the experiments and have fun with them. If you want to do more, there are lots to try on the internet. You can learn more about convection currents, thermodynamics, clouds, rainbows and thunderstorms. You can even make your own weather recording devices!

If you do try any weather experiments, be sure to send them to me at Met Éireann at joanna.donnelly@met.ie. I love hearing from the public about the weather! Met Éireann also has a new website at www. met.ie. Here you can find all the latest weather for your area and the whole country, along with lots of interesting meteorological information and history.

When I was growing up, one of my favourite places to visit was our local library. I still remember the little blue card that was my ticket to the magical world of books, and like Roald Dahl's Matilda, I worked steadily through the children's section until they finally let me through the door to the adult books. The library is still one of my favourite places to go, both by myself and with my kids. You can find answers to everything in a library.

These days, we also have the internet. Every question we can think of can be answered almost instantly with a device we can fit in the palm of our hand. But that can slow us down too. If we aren't exercising our brain, it can get sloppy and tired, just like when we don't exercise our bodies. Remember, just because something is online, doesn't mean it's true. Learning all the facts and understanding them yourself is the best way to keep your brain in shape.

The next time someone tells you something silly, like that there's gold at the end of the rainbow, or that climate change isn't real, you can tell them the truth. That way, the fact will have gone from me to you to someone else, and they might tell another person, and another, and it might even go all the way around the world.

So don't stop asking questions. There is no such thing as a silly question – the only silly questions are the ones we don't ask. And at the end of the day, that's what science is about: asking a question, then finding the answer.

But the funny thing about science is that the more you know, the more you realise you don't know. Once you find the answer to one question, there's always another hiding behind it. Now you know why the sky is blue. But do you know why the sea is salty? If you don't, you know what to do – ask another question!

And finally, we all need to look after the planet. We've only found one so far with the right atmosphere for plants, animals and humans to live and grow.

Remember what we learned earlier – in weather forecasting, small differences can add up to big effects. It's exactly the same with people all over the planet. We need to make small changes to our lifestyle to make the big change that the earth needs. We're all butterflies, and if we flap our wings together, we might just make a hurricane!

Good luck.

Joanna

INDEX

ABOUT THE ILLUSTRATOR & AUTHOR

FUCHSIA MACAREE is an illustrator from the border of Clare and Tipperary, now living and working in Dublin. She graduated from Visual Communication in NCAD in 2011, followed by an MA in Illustration at the University of the Arts, London. She is interested in human interactions, Irish oddities, humour, simplified shapes and strong colours. She shares an art studio in the city centre with her friends where she makes books, posters, murals and maps. She loves sea swimming and cycling. Her favourite type of weather is a crisp autumn day for kicking leaves, but she also loves being cosy inside while hearing rain hit the roof.

JOANNA DONNELLY is from Dublin and joined Met Éireann in 1995, where she has worked as a meteorologist, forecaster and broadcaster. Joanna is married to fellow forecaster, Harm Luijkx, from The Netherlands. They have a daughter, Nicci, and two sons, Tobias and Casper. Joanna is very active and passionate about the environment so she cycles 16 kilometres to work and grows food in her garden. A nice day for her is sunshine and showers as it's best for the plants, but also for seeing rainbows! Her favourite weather of all is a good thunderstorm, because they're a wonderful example of the power of nature.

GILL BOOKS

Hume Avenue
Park West
Dublin 12
www.gillbooks.ie

Gill Books is an imprint of M.H. Gill and Co.

Text © Joanna Donnelly 2018
Illustrations © Fuchsia MacAree 2018

978 07171 8093 6

Proofread by Ellen Christie
Indexed by Eileen O'Neill

Printed by L.E.G.O. SpA, Italy
This book is typeset in ITC Century Book, 15.5 on 21 pt.

The paper used in this book comes from the wood pulp of managed
forests. For every tree felled, at least one tree is planted, thereby
renewing natural resources.

A CIP catalogue record for this book is available from
the British Library.

5 4 3 2 1